Fair Food

Stories From A Movement Changing The World
Edited By Nick Rose

16
EasyRead Large

RHYW

Copyright Page from the Original Book

First published in 2015 by University of Queensland Press
PO Box 6042, St Lucia, Queensland 4067 Australia

www.uqp.com.au
www.australianfoodsovereigntyalliance.org

Cataloguing-in-Publication entry is available from the
National Library of Australia
http://catalogue.nla.gov.au/

ISBN (pbk) 978 0 7022 5366 9
ISBN (ePdf) 978 0 7022 5542 7
ISBN (ePUB) 978 0 7022 5543 4
ISBN (kindle) 978 0 7022 5544 1

Edited by Nikola Lusk
Typeset in 11.5/16 pt Bembo by Post Pre-press Group, Brisbane
Printed in Australia by McPherson's Printing Group

University of Queensland Press uses papers that are natural, renewable and recyclable products made from wood grown in sustainable forests. The logging and manufacturing processes conform to the environmental regulations of the country of origin.

TABLE OF CONTENTS

Dr Nick Rose is the National Coordinator of the Australian Food Sovereignty Alliance, and Food Systems Coordinator of a new National Food Network, 'Sustain'. He was one of the developers of Australia's first crowdsourced food policy document, the People's Food Plan, and jointly coordinates Fair Food Week, now in its third year. He was the Content Director of Australia's first food politics documentary, *Fair Food.* Nick was awarded his PhD in Political Ecology from RMIT University for investigating the transformative potential of the global Food Sovereignty movement. As a Churchill Fellow, he has presented his findings on urban agriculture to a wide range of audiences in Australia and continues to advocate for the adoption of the recommendations.

FOREWORD

DAVID POCOCK

This is a book about food. But it's also a book about people. About remarkable folks who've made connections between what we eat, how it's grown and what our future, as a species and as a planet, might hold. It's a book about the potential for change and the power of direct action. In a world seemingly full of disaster, it's a book about hopeful futures, about community resilience.

In our contemporary Western world, we are bombarded with statistics. They generally paint a rather bleak view of things, particularly statistics related to the food system, whether it's food waste, food miles, environmental degradation, climate change, obesity and type 2 diabetes, hunger and malnutrition – the list goes on and on. From all the statistics it seems pretty clear that our food system is broken. Thinking like this is not only daunting, it's problematic, because when we talk only about statistics we lose sight of the individuals who make up those statistics; they become an abstract number. And in the face of such harrowing statistics, it is these stories we desperately need: the stories of people who are fighting against the current food system and providing alternative visions and practical solutions.

After reading *Fair Food,* I was reminded of the Nigerian poet and novelist Ben Okri, who said: '*Stories* can conquer fear, you know. They can make the heart bigger.' The stories in this book are of people who have not allowed fear to stop them from working towards a different food system, a different way of relating to the earth. These are stories of people with big hearts who've toiled to care for their soil and their communities.

We each have our own stories, and a move to a saner, fairer food system can be part of all our stories. Indeed, if we are to move to a Fair Food system, it may take all of us. My own story with food began at birth, even before that. I was born as the third generation in a farming family. Both sets of my grandparents farmed in Zimbabwe – on my mum's side as citrus farmers and conservationists; on my dad's as cattle farmers, vegetable growers and flower exporters. My childhood years were spent on the farm and in the African bush. These were years full of the kind of learning one never realises is happening as a child. Learning about different varieties of tomatoes, listening to my dad and Phanuel, the foreman, discussing cattle management, watching Mum manage the processing of hundreds of chickens and witnessing a hailstorm destroy every crop on the farm – learning how farmers are truly at the mercy of the climate.

By the time I was a teenager, my family, through circumstances beyond our control, had been dislocated from the land, and we found ourselves in suburban

Australia. It took me many years to come back to the land, but in my early 20s I found myself yearning for more of a connection with nature and a connection to my own food, with soil under my fingernails, sweat and hard work.

This yearning came at the same time as I was beginning to think more and more about issues of injustice in the world, issues that seem very big and often hard to engage in – yet many of them play out in our food system, something we all interact with on a daily basis.

Growing up on a family farm and seeing how our prospects depended not only on the land and the weather, but also on the market – and factors well out of our control, sometimes on the other side of the planet – gave me a sense of the complexity of concerns faced by food producers and, therefore, by all of us. It is one of the main reasons I'm so interested in and committed to supporting good farmers – farmers who not only produce valuable goods, but also look after the land, conserve soil and create a localised and more ecologically sustainable food system.

The stories in this book are just a few of the many thousands of stories of people engaging more and more meaningfully with the food they eat – from market gardeners to livestock producers to backyard gardeners and people who are buying direct from growers at farmers' markets and through community-supported agriculture, to shop owners, butchers and wholesalers who are finding ways to make things fairer for producers, the earth and animals. We all have a role to play in the future of Fair Food. Its time has come.

I hope that this book and the stories in it will inspire you to think more about the food you eat, the farmer who grew it, the land it grew on and how you might contribute to making the food system more fair. *Fair Food* tells of the future that is possible, the future that has already begun on farms, in paddocks, in markets and on plates all around Australia. When all of our stories become connected, the tide of Fair Food will be unstoppable. In the words of Tammi Jonas, pig-farming butcher and one of the authors featured in this book, 'Watch out, Big Food, because we are legion, and the future of Fair Food is now.'

David Pocock

FOREWORD

GUY GROSSI

My father was a chef and was passionate about growing things. He was raised in a small town in the south of Italy called Carosino in the 1930s and 1940s. It was common for people of this town to have small landholdings, which they used for growing a variety of fruit and vegetables like the famed Primitivo – a grape variety used for making the local wine – tomatoes, eggplants and more, as well as olives for oil. It was also common to exchange produce in a kind of community-sharing system so that more variety could be had. Many of these practices existed because of economic pressures. I always enjoyed listening to my dad's stories about the customs of his childhood, how he was taught about the dirt and how their small piece of land needed to be cared for as it helped to feed the family.

I was fortunate growing up in the suburbs of Melbourne. We enjoyed fresh seasonal fruit and vegetables on a regular basis straight from the backyard garden. It was the early 1970s and cultural diversity was growing fast. In many ways food became a conduit and helped in a sometimes challenging social environment. Tomato day was the norm each year, and spending time on a Sunday in the garden with my family was just what you did before being summoned for a lavish and tasty lunch.

I was 15 when I started my apprenticeship, though I'd had some experience in commercial kitchens accompanying my father to his work over the years. The focus for the newcomers in the kitchen was simple: work hard and cook from the heart to produce a high-quality product that would keep guests happy and coming back for more. There was talk about great produce and of course consideration given to provenance, but, looking back, it was more about marketing than real quality. We were always keen to find out more about what we were cooking, but were we asking all the right questions?

By the time I had my children, the larger food sellers were replacing many smaller retailers, so butchers, greengrocers and many others were fading from the street fronts. I had no backyard garden! The 'norm' had changed and, without realising it, I was becoming part of the increasing and serious problem that is damaging our food system. I have always believed a person in business has a duty to evolve, and standing still is the enemy of great practice. A few years ago, however, some events in my life caused me to look at things much more deeply. My view of this topic got clearer. It was like opening a window, and what I saw was how much more I wanted and needed to know about how our food comes to us and how it is produced. It really is a shift in thinking, and connecting with others on this has had a profound impact on me.

I met Nick Rose when we were organising the first Melbourne Tomato Festival, which took place at Farm Vigano in South Morang on 1 March 2015. He has an incredible past and has done so much in this area of Fair Food. His enthusiasm is infectious and leaves you starving to know more. He came on board to help and to be part of a panel discussion at the festival itself called 'Every ingredient is sacred'. The talk was a full house on the day. It inspired quite a stimulating discussion on the importance of this issue, now and for the future.

As a human, maintaining my family traditions is becoming more important as I mature. The Tomato Festival itself is part of a movement that we have coined the 'Melbournese Movement'. Its purpose is to promote the preserving of tradition, culture and the connection between food and social behaviour and our community.

As a chef, I need to know where and how the food I buy is farmed. I need to know about the quality of the food I am using. Quality is a more defined word to me now and means more than what has ended up on the workbench; I also need to know how it got there. Keeping food as local as we can, when we can, and knowing what we choose to do has a great effect on others.

In this book, Nick has brought together 10 authors each with their own personal stories of cultural change and transformation.

It is by sharing these stories and connections that change can be made. This wonderful compilation will clearly show the impact on our earth and provide deep and insightful understanding of what needs to change. In Nick's own words, 'We can all become agents of change.'

And I have my own backyard garden now!

Guy Grossi

INTRODUCTION

NICK ROSE

This is the text of the speech I deliver when introducing the *Fair Food* documentary, which first premiered to a sell-out audience at the National Gallery of Victoria on 2 December 2014, and has since been screened at dozens of community venues across Australia. It serves equally well to introduce this book, and I have adapted it accordingly.

***

Why did we write this book? Because the food system is broken.

Why is it broken?

Because we have fully applied the technologies and the mindset of industrialisation to food and farming. And because we have combined industrialisation with the logic and the imperative of endlessly increasing production, regardless of the consequences.

What does that mean? It means we have over-exploited our land, degraded our soils and damaged our river systems. It means Australia has one of the highest rates of deforestation, biodiversity loss and species extinction on the planet. It means, globally, that the food system contributes as much as 50 per cent of all greenhouse gas emissions.

It means that we have a supermarket duopoly that controls 70 to 80 per cent of the grocery market, forcing farmers and food processors to take whatever prices the duopoly offers. A hundred years ago farmers received 90 cents of every dollar's worth of food they produced; today it's around 10 cents.

Farming has become devalued in our highly urbanised culture, and not just economically. So it's shocking, but not surprising, that seven farmers leave the land every day, and that rates of suicide and depression among farmers are twice the national average.

Our industrialised food system produces too much food of the wrong type. So we're subjected to an endless barrage of advertising, urging us to buy food products laced with excess sugars and salt. Dietary-related diseases are already among the biggest public health issues we face.

Our food system is not merely broken. It's *killing* us, and ruining any chance that future generations have for a decent and livable future. Yet the industrialised food system persists, and is expanding. Why? Because there are very powerful economic and financial interests that make a lot of money from the status quo. Because we, the people, are so disconnected from our food system. Because food is apparently abundant and cheap, and because we don't join the dots of what all of this means for us in the long term.

We wrote this book, and we formed the Australian Food Sovereignty Alliance, because we can no longer tolerate this state of affairs. Because it's *no longer enough* just to talk or think in terms of reforms. We need a transformation; we need a *revolution.* And that revolution begins in our own minds, in our hearts, in our consciousness.

This is the challenge to every one of you reading this book. This is the choice facing every one of us alive today. Do we continue to allow our culture and our society to become ever more destructive, and ever more violent? Do we choose to remain in a paradigm that says that the earth, and indeed ourselves, only exists for endless exploitation so that a tiny fraction of humanity can enjoy obscene levels of wealth?

Or do we choose to be part of the great challenge of our times – the *greatest challenge of all times?* To create a *shared vision* of a wonderful, bountiful world, where there is no hunger and no poverty; where soils are thriving, rivers are healthy and forests are abundant; where animals roam freely; and where all of us are healthy and flourishing.

Do we choose to see ourselves as victims of processes and powers beyond our control, and simply walk away and do nothing, resigned to our fate? Or do we choose to see ourselves as subjects and shapers of our own history, as creators and narrators of our own story, as powerful beings with the capacity to effect great changes?

Because I'm here to tell you, *that's* who we are. We *are* powerful. In the words of the thirteenth-century Sufi mystic and poet Rumi:

You were born with potential.

You were born with goodness and trust.

You were born with ideals and dreams.

You were born with greatness.

You were born with wings.

You are not meant for crawling, so don't.

You have wings.

Learn to use them and fly!

We wrote this book because these are messages that need to be heard. This is the story that *needs* to be told, that we need to tell ourselves, and each other. We wrote this book because we know that there are women and men all over this country, and all around the world, who have embraced this new paradigm, who are blazing a trail towards the decent, fair and livable future that all of us want.

We wrote this book to recognise and celebrate them – and ourselves. They – we – are our Fair Food pioneers. And this is the story of Fair Food.

FROM CORPORATE LAWYER TO FAIR FOOD SCHOLAR-ACTIVIST

NICK ROSE

We might describe the challenge before us by the following sentence. The historical mission of our times is to reinvent the human – at the species level, with critical reflection, within the community of life-systems, in a time-developmental context, by means of story and shared dream experience.

Thomas Berry, *The Great Work*

AN EARLY EPIPHANY

I was 13 years old when I broke my neck. I was sitting in the back seat of the family Range Rover, asleep. We were 25 days into a five-week, 5000-kilometre family 'holiday', driving from Perth to Kalgoorlie to Uluru, across to the Simpson Desert then down to Adelaide and back across the Nullarbor Plain. For a 13-year-old, it was a seemingly endless string of eight-hour days driving through mostly featureless rocky desert, with nothing but the collected works of Slim Dusty on 50 cassettes for distraction.

What was the point of that huge trip? It seemed to be to go as far as possible in as short a time as possible. To devise a schedule, a list of towns and sites to visit and plan the routes by which to get there, and then to follow the schedule exactly. To avoid having conversations about things that mattered. Oh, and to collect stickers and pennants of the places we visited. I collected the stickers, my younger brother the pennants. For a while they graced the walls of our respective bedrooms, and then, when we left home, they hung in the corridor of my parents' house. A sombre memento to a near-death experience.

And maybe, just maybe, the whole trip serves as a metaphor for the modern globalised, industrial economy. To go as far as possible – to accumulate as much money and wealth as possible – in as short a possible time. To collect many useless trinkets along the way. To skim over the surface of life without pausing to consider its beauty and complexity, to really *see* it, and *know* it. To avoid reflecting, and asking where we are going, and why.

Perhaps, as a species, we need to have a near-death experience. Something that rouses us from our lethargy, reminds us of who we are and what we are supposed to be doing. Perhaps that's exactly what we're having now, in the converging crises we're experiencing. Perhaps our wake-up call has arrived – if we choose to see it.

On this particular day, in August 1981, it must have been early afternoon when my sister, Gillian, was driving along a rocky gravel road, and the car got into a fish-tail spin, as Range Rovers of that era were wont to do. Some memories from childhood are indelibly imprinted in the mind's eye, and this is one of them. I was sitting behind the driver's seat, asleep, and not wearing a seatbelt. I opened my eyes briefly, saw that the car was swerving all over the road, heard my mum say to Gillian, 'Don't brake!' and had a vague sense of the panic that was gripping the car. Then I promptly closed my eyes and went back to sleep.

Ever since I have wondered why I did that. In all the thousands of kilometres I've travelled since, using various forms of road transport in many places around the world, I don't think I've ever really fallen asleep once. And if a car or a bus was about to run off the road or have a crash, I can't imagine I'd do anything other than sit bolt upright, terrified, and prepare myself for impact.

But not that day in August 1981. I registered what was happening and went straight back to sleep. Almost as if someone was saying, 'Relax, it'll be alright. Stay asleep.'

The next thing I remember is waking up, on my back, on the ground, with the car some distance away. I glanced at it and saw it was half-squashed, the roof mostly caved in, like an empty tin can. The other occupants – my mum, my aunt June, Gillian, and my younger brother, Justin – were also strewn across the road. I lifted my right arm, which was hurting, and saw the wrist distorted in an unnatural form. The pain got worse. I felt some numbness, a dull ache, in my neck.

How long we all lay there, I have no idea. At some point a car passed and alerted the medics in the nearest town, thankfully only about ten kilometres away. An ambulance came and collected us.

I remember the GP inspecting me. He said, 'I can see you've broken your wrist. Are you hurting anywhere else?' To which I replied, 'Yes, my neck's a bit sore.' He felt around my neck for a little while and then said, 'That seems fine.' I was told to get to my feet, walk out to the waiting ambulance, which took us all to the local airfield. We spent the next two hours in a light aircraft, which flew us to the nearest town with a major hospital, Whyalla.

I was sitting up all that time, while Gillian, whom the doctor suspected had whiplash (as it turned out, she didn't), was lying down on a stretcher. In Whyalla, X-rays of my neck were taken, and they discovered that my fourth and fifth vertebrae were both fractured, with significant tendon and ligament damage as well. I spent the next six months wearing a foam neck brace, before they performed a fusion operation in Darwin, taking a slice of bone from my hip and fusing my fourth and fifth vertebrae together. That was a precautionary measure, to avert the possibility of further harm should I play contact sport, which I later did for a brief period.

Thirty-four years later, I am still walking around quite happily, and have two large and impressive scars to use as props whenever I tell this story. But the question I still have is this: why did I survive that day? I could so easily have died in that car crash. I could so easily have moved the wrong way, or the light plane could have got into some nasty turbulence, when I was upright with an unsupported broken neck, and have become a quadriplegic.

And the most pressing question is this: *why did I go back to sleep?* I have become convinced that had I been rigid, I would likely not have survived.

Ultimately I will never know for sure, but what I believe is this: I didn't die that day in 1981 because I had work to do. My life had a purpose; I was here for a reason. Of course, I didn't know at the time what that purpose and that reason was. It's taken me a long time to find out. But I think now that I have.

That's what this book is about. We all have moments like these in our lives, turning points, big decision times, epiphanies perhaps; moments where if we say 'yes' instead of 'no' it can have repercussions that can last years, decades, even generations. Me breaking my neck might be more dramatic than some and less dramatic than many others, but these are the moments that add depth and richness to the fabric, to the story, of our lives. These are the moments where, either at the time itself or (more usually) when we look back, we can see our life taking a certain course. Steering us in a certain direction.

These are the moments when the meaning of our lives can, if we choose to see it, become more clearly visible to us. They are the moments when the contours of our own stories start to take form and shape.

SHARING STORIES

This is a book of sharing and reflection. I want to share with you some important parts of my own story and those of a number of friends and colleagues. These are individuals whom I have met along the path of my own journey to becoming a Fair Food scholar-activist. Each in their own way has made, and is making, a major contribution to a fairer and more sustainable food and farming system for Australia.

I am not claiming that the stories of these individuals – and much less my own story – are the most important stories to be told. There are thousands – if not tens of thousands – of people around the country whom I would describe as 'Fair Food pioneers': women and men who are striving in their own places, working with their communities, using the skills and knowledge they already possess and acquiring new capacities to create something better than what currently exists. Each in their own way is driven by a larger vision of that better food system, that better world.

The reason I have invited the nine individuals whom you will meet in the succeeding chapters to tell their stories in this book is because my encounters with these people have become an important part of my own story. I feel formed, in a very real way, by the relationships I have in my life. Isn't that true of us all? When we're asked, 'What is most important to you in life? What gives your life meaning and purpose?' we normally answer, 'family and friends'. Indeed. Where would we be without them? *Who* would we be without them? *How* would we be without them?

So these 10 individuals are, in a very real sense, *part of me.* They are woven into the fabric of the story of my own life. Some I have known for quite some time, some I have met more recently. All have left their imprint on me. All are remarkable individuals in their own right.

But there's more to it than that. Each of these individuals has, like me, gone through moments of profound personal and psychological change. Each has made choices with lasting consequences. Each has experienced an epiphany, we might even say a calling or a vocation, to embark upon what the great American agrarian philosopher Wendell Berry, in his book *What Matters?,* describes as work that serves 'natural and human communities, not arbitrarily "created" jobs that serve only the economy'. Work that supports human flourishing, in all its richness, and in harmony with the living world of which we form part.

This kind of work is part of what cultural historian Thomas Berry describes as the 'Great Work': 'to carry out the transition from a period of human devastation of the Earth to a period when humans would be present to the planet in a mutually beneficial manner'. I would say that every single one of us has a calling to do 'work that matters', and play our role in the 'Great Work'. What it is for each of us personally, only we can know and discover. But how many of us actually feel we have the opportunity to live our vocation? How many of us are performing jobs that don't *feel* right – but that we do out of necessity, or habit? How many of us don't have 'jobs' at all, so that we feel marginalised and devalued in this very materialistic and often ruthless culture?

Each of the stories you will read describes how the individuals in question came to discover their vocation: what experiences, encounters, traumas, conversations and/or texts led them to realise that they too had important work to do. And each of these individuals is, like me, enmeshed in a large and expanding web of relationships and community; all of us share a strong conviction that major change is needed right now, and that we have the capacity, individually and collectively, to help bring that change about.

That's why these stories are part of a much larger story. It's the story of transforming the food system – and through the food system, our culture, our economy, our politics and our society. As we transform ourselves, so we begin to transform the world.

These encounters, the forming of these types of relationships, are happening all around the world. As more of us discover – or remember – our vocation, we are finding each other. Connections are multiplying exponentially.

The big change, the great transition, from one paradigm to another, is well underway. It's from a paradigm of separation and disconnection to a paradigm of healing and connection. Can you feel it?

LEAVING RIALTO TOWERS FOR A TRAVELLING LIFE

I was a rebellious and angry teenager, but ultimately a dutiful and loyal son. My parents worked hard and sacrificed much so that their six children could have the 'best education that money could buy' (in Perth, at least) and so (in theory) embark on the road to a professional career and 'success'. In my parents' vision, that meant becoming a doctor or a lawyer. I was the fifth child, and, for various reasons, none of my elder siblings had fulfilled the family dream of becoming a professional. I studied hard in my last year at high school, and gained entry to Melbourne

University Law School for 1985. Off I went, a bright-eyed 17-year-old.

Somehow I got through the four years of a law degree, mastering (after a fashion) but not really *understanding* subjects like 'property' and 'contract' and 'employment law'. It was only much later that I began to appreciate just how fundamental these concepts and institutions are to so much of our culture and society. The way law was taught in the mid 1980s, there was never any historical contextualisation of how and why, for example, the institution of 'private property' came into existence. Nothing was ever said about the foundational role that dispossession, deceit and violence – at home and abroad – played in building the 'civilisation' and the 'economy' we inherit today. These basic historical facts were simply airbrushed from the legal texts used to form the minds of the new generation of barristers and solicitors, whose role it is to further solidify the institutions of property and contract in pursuit of the taken-for-granted assumption that our main collective goal is to secure the endless expansion of the market economy (otherwise known as 'economic growth').

So in 1989 I found myself as an articled clerk at Mallesons Stephen Jaques, working in the southern Rialto Tower in Collins Street, Melbourne. I had just turned 21. A year later, I was admitted as a barrister and solicitor of the Supreme Court of Victoria. I had my own office with floor-to-ceiling glass, on the 45th floor, in the industrial relations and employment law department. I had a secretary to take my calls and type the letters that I dictated. I was taken out for lunch and drinks by keen young barristers who wanted me to send them briefs. I had a taste of the trappings of wealth and status, and was on the lower rungs of a ladder that, had I continued climbing and met the demanding performance targets ('billable hours' was how productivity was defined – and no doubt still is), would have seen me reach the upper few per cent of income earners in the country in perhaps 10 to 15 years.

But though I could do the work, and even enjoyed some of it, being a corporate lawyer never felt right to me. I felt too young, too immature, as though I hadn't actually lived. I felt somewhat fraudulent – who was I to be advising business managers on things as momentous as firing employees for misconduct, or making workers redundant or embarking on litigation? I might have known and been able to interpret the relevant legislation and case law, but what life experience could I offer in such situations? How was my advice ever going to be 'practical'?

Still, I stuck at it for the best part of two years, post-articles. And the reason I left was not primarily my own moral qualms or sense of inadequacy. Rather, it was because I fell in love, and pursuing that relationship became far more important to me than pursuing a career as a corporate lawyer. The story of how Julie – with whom I have travelled half the world and had two sons, who has supported me through a

PhD and shared dozens of tribulations in my years of angst and self-doubt, and to whom therefore I owe an enormous amount of gratitude – came to meet and be together with me is again revealing of how circumstances seem to conspire to bring certain people together at certain times, for reasons that often only become apparent later.

Julie grew up in Sydney, and finished high school and studied law at the University of Adelaide. She never wanted to be a lawyer, and went to work for Shell at the Geelong oil refinery as an industrial relations officer. It was only after nearly four years in the job, when a move back to head office in Melbourne was imminent, that she seriously thought about embarking on a career as a solicitor. And she wanted to work for Freehills, the firm that Shell used, not Mallesons. Indeed, she had grave misgivings about the industrial relations department at Mallesons, having eyeballed the four partners who headed it up in her job interview. But there was no job going at Freehills at the time, whereas Mallesons welcomed her with arms open.

Julie was engaged to be married when we met for the first time towards the end of 1989. She and her fiancé owned a house in Carlton. Their lives were mapped out, clearly signposted: wedding, children, career, bigger house, holidays ... I was also in a relationship that I'd started in my final year at uni, but even though we were living together there were no wedding plans, no house-buying plans. It took a little while for the office romance between Julie and me to flourish, but when it did there was no holding us back. Corporate lawyers are notorious for working long hours, so we had the perfect excuse to go out to city pubs, drink wine and smoke lots of cigarettes, which you could still do inside pubs in those days. About halfway through 1990 we ended our respective relationships, and by the end of that year we were living together in a share house in Parkville.

We dreamt of travelling, which neither of us had done. We dreamt of freedom, of not having to get up and put on a suit and commute to the office. Of no longer having to perform the role of corporate lawyer. We bought Lonely Planet books. We made plans. By August 1991 we had resigned, and by November we had left Australia. We thought we might be away 18 months. Over the next 15 years, we only came back once, for three weeks, in 1999.

That was one of those choices when, in saying 'yes', our paths took an utterly different direction.

FOOD SOVEREIGNTY: THE WORLD'S LARGEST SOCIAL MOVEMENT

The stories contained in this book together form part of a much larger story. As I said earlier, it's a story of healing and reconnection. It's a story of culture change, system change: of transformation. According to some writers and thinkers, such as Thomas Berry and cell biologist Bruce Lipton, it might even be a story of a qualitative leap in our own evolution as a species.

How does that leap happen? Very briefly, because our culture, our economy and our society are no longer congruent either with our levels of scientific understanding of the universe or with indigenous cosmologies and ways of knowing that see the world and all its people as fundamentally interconnected and interdependent. With Einstein's discoveries and the insights revealed by quantum physics, as well as social network theory, social psychology and many other disciplines, the idea of us being completely separate and atomised individuals, who have an innate drive to maximise our own self-interest even at the expense of others around us, and indeed at the expense of the living planet, appears both dangerously quaint and absurd. Yet many of us continue to live our lives as though that idea were an established and unalterable 'fact'. It's not. On the contrary, it's an ideology, and it serves very definite political and economic interests. The challenge we now face is to transform our understanding of ourselves and our culture, so that our worldview and our institutions recognise *collaboration* and *cooperation,* rather than self-interested *competition,* as the primary driving forces of our lives, individually and as a species.

This book is about one movement in particular, which seeks to embody and narrate this new cultural paradigm: the global movement for Food Sovereignty, or 'Fair Food' as we have called it in Australia. It's important to be clear: while everyone in this book is working for a fairer food system, we're actually working for a transformed economy and society as a whole. Food is where we are concentrating our energies, because it is so fundamental to so much of who we are and what we do; our food system is extraordinarily destructive in social and environmental terms, which means a transformed food system can have extraordinary healing and restorative powers. Yet food is only one of the big systems that require transformation, along with energy, transport, housing, education, health and ultimately our economy and political systems as a whole.

Food Sovereignty is a new movement, barely 20 years old. It emerged in the mid 1990s out of cultural exchanges between farmers in North and Central America and Europe, which took place as the 'free trade' era in agriculture was being ushered in under the auspices of the World Trade Organization. These exchanges led, in 1993, to the formation of La Via Campesina ('The Farmers' Way') as a global social movement foregrounding the voices and experiences of the hundreds of millions of small and family farmers who grow food primarily for themselves and their own communities. Food Sovereignty embraces over 300 million people in 80 countries around the world, which makes it quite likely the largest contemporary social movement.

In my own research, I describe Food Sovereignty as being founded in an 'ontology of connectedness', which contrasts with the dominant, globalising and industrialised food system that I suggest is based in an 'ontology of separation and alienation'. 'Ontology' refers to a 'theory of reality': it asks, how do we understand the world around us, and our place in it? So by 'ontology of connectedness', I am suggesting that Food Sovereignty is both a theory and a practice in which increasingly personal and intimate relationships are formed around food. A classic example is farmers' markets, where you can buy 'food with a face' directly from the person who grew or raised the food (in the case of true producers' markets). Even more direct is buying from the farmer, at the farm gate or farm shop. Community-supported agriculture is another expression of a connected food system, where the farmer (or group of farmers) sells shares or subscriptions in a growing season, or animal or herd, to a group of nearby residents.

Most direct – most connected – of all, of course, is the food you grow yourself. And millions of Australians are doing exactly that – more than half, according to recent survey research from the Australia Institute. We *want* to be more connected to our food. And we are taking steps to achieve that outcome.

MY POLITICAL AWAKENING

After 16 months of backpacking through south-east Asia, India and Nepal, Julie and I found ourselves in Kathmandu on a freezing January day in 1993, riding away in a rickshaw from the British High Commission. Our savings had pretty much run out and we were deciding whether to continue on to England or return to Australia. We wanted to keep going, and I had even shaved off my dreadlocks that I'd proudly managed to grow over the previous six months travelling through India, all to make a good impression on the consular staff. To no avail: while Julie had a British passport (her parents are from Yorkshire), I was told bluntly that I would need $4000 to enter the country on a working holiday visa.

What is Food Sovereignty?

According to the first global peoples' Forum for Food Sovereignty, which took place in Mali, in February 2007, Food Sovereignty:

- gives people the right to healthy and culturally appropriate food produced through ecologically sound and sustainable methods, and to define their own food and agriculture systems

- puts the aspirations and needs of those who produce, distribute and consume food at the heart of food systems and policies rather than the demands of markets and corporations

- defends the interests and inclusion of the next generation
- offers a strategy to resist and dismantle the current corporate trade and food regime, and directions for food, farming, pastoral and fisheries systems determined by local producers and users
- prioritises local and national economies and markets and empowers peasant and family farmer-driven agriculture, artisanal-fishing, pastoralist-led grazing, and food production, distribution and consumption based on environmental, social and economic sustainability
- promotes transparent trade that guarantees just incomes to all peoples as well as the rights of consumers to control their food and nutrition
- ensures that the rights to use and manage lands, territories, waters, seeds, livestock and biodiversity are in the hands of those of us who produce food
- implies new social relations free of oppression and inequality between men and women, peoples, racial groups, social and economic classes and generations.

In other words, Food Sovereignty is a vision and political program for a transformed society.

We decided to ignore the consular official's advice, and a couple of weeks later arrived at Heathrow. I was half expecting to be turned around, but instead was welcomed with a smile and a two-year stamp in the passport. I was learning that calculated risk-taking and listening to intuition were good things to do.

The first year was hard. We had little work, and lived in a freezing flat in Manchester. One September night, while serving pints of lager in the Horse & Jockey on Chorlton Green, I was punched in the face for my troubles. All for the princely sum of $6 an hour plus some meagre tips. What a come-down from the 45th floor of the Rialto Tower. I was desperate to take any job, and it seemed at that point that we would indeed go back to Australia, far poorer than when we left.

Even then I had not abandoned ambitions of corporate law. But it was a recession in England, and law firms weren't hiring junior Australian solicitors. Just when it all seemed hopeless, I applied for and got a job as a writer for the London-based publishing company Incomes Data Services (IDS). The editor took a punt on me, and it meant we could move to London. The following August we bought a two-bedroom flat in Kentish Town. I left IDS in 1995 to work for a newly formed charity, Public Concern at Work, which was providing legal advice to employee whistleblowers and campaigning for law reform to secure for them greater protections against unfair dismissal and harassment. It was there I got a taste for campaigning and advocacy work.

In 1997, I left Public Concern at Work and joined the legal department of one of Britain's larger trade unions, Manufacturing, Science & Finance (MSF, now Unite). By then I knew clearly my sympathies lay very strongly with working people, so being an employee of a trade union seemed a good fit. And I was quite happy there for three years, but what I also learnt in those seven years in London is that I don't do offices very well. I just don't think they suit my style of working. And in MSF there was something of a culture of mediocrity, where people had been in the same jobs for years, in some cases for decades. The passion and the energy weren't really there.

Comfortable though we were, something was pushing Julie and me on to a new challenge. In 1999, Julie had found out about Peace Brigades International (PBI), a Spanish NGO that provided 'human rights accompaniers', or 'unarmed bodyguards', as they were sometimes called, to work and/or live alongside human rights defenders in countries experiencing internal conflicts. The theory was that by placing foreigners in such situations, it would – with the appropriate diplomatic safeguards – act as a deterrent to those making threats of violence and intimidation from actually carrying out their threats. PBI started in 1984 with a mission to Guatemala, which was at that time emerging from the horrors of the worst five years of a long and brutal 36-year internal armed conflict. By 1999, the main project was in Colombia.

The idea of human rights accompaniment appealed to us. We were – and are – idealists. This would be a challenge: to learn a new language, to live in a new country, to understand a new culture, to work alongside brave people facing threats of violence because of the work they did to protect and benefit others.

First, we had to learn Spanish. So in early 2000 we quit our jobs, put our personal things into a storage locker, rented our flat and headed to Guatemala via the United States and Mexico for some weeks' tuition in a Spanish language school in the north-west highlands city of Quetzaltenango, or Xela, as it's often called.

The year 2000 was probably the most critical year for me in finally shedding any illusions of being a corporate lawyer, and starting to embrace a new identity. I wasn't sure what that was, but I knew it had to be something political. Something that challenged entrenched power and privilege. Because after just a few weeks living in Xela, and attending the language school run by sympathisers of the ex-guerrilla army – *Proyecto Linguistico Quetzalteco de Español* – I quickly came to appreciate the tragedy that the poor, largely indigenous, majority of that Central American country had endured for decades. And are still enduring.

I discovered, in the school's excellent library, a 1982 book written by Stephen Schlesinger, Steven Kinzer, Richard Nuccio and John Coatsworth: *Bitter Fruit: The Story of the American Coup in Guatemala.* Reading this while in Guatemala, seeing the entrenched levels of poverty, the routine daily violence, and having conversations with Guatemalans about all these issues, had a profound impact on me. Academic historians would no doubt call what I am about to say a gross over-simplification, but my takeaway message from that book was this: the Eisenhower administration deliberately organised a coup d'etat in 1954 to overthrow the popular social democrat president of Guatemala, Jacobo Árbenz, essentially because he dared to intrude upon the private property rights of a US corporation, the United Fruit Company (UFC).

Never mind that UFC owned most of the best arable land in the country, as well as the railroad and the ports, and kept the indigenous majority eking out a bare subsistence living on substandard small patches of land on hillsides. It was when Árbenz decided to expropriate some of UFC's landholdings, to redistribute them to the *campesinos,* and compensate UFC on the value the company itself had declared for tax purposes (always low, because UFC preferred not to pay much tax), that the CIA and the White House decided enough was enough. Árbenz was assassinated by hired guns, a military dictatorship was installed, and in 1960 a brutal armed conflict began that would leave over 200,000 dead, tens of thousands more tortured and disappeared, and which culminated in a genocide against four Mayan ethnic groups in the period from 1978 to 1983.

Julie and I lived in Guatemala for the next six years. We got some work in the field of human rights capacity building, assisting Guatemalan grassroots organisations to become more effective in their own communities, and helping build networks among organisations across Guatemala, Honduras and Chiapas. Our two boys, Camilo and Jude, were both born there, in 2001 and 2005. In 2003, I returned to study, starting a Master's in International and Community Development; with the advent of the internet I could do it all from Guatemala. I became familiar with critical theory; I learnt about the Howard government's demonisation of refugees; I informed myself in depth about gender inequality; I studied and wrote about the War on Terror. My political consciousness was now well and truly awakened, but yet to find an outlet. As passionate as I felt about the situation in Guatemala, I realised that it was not 'my' country and ultimately not 'my' struggle.

By August 2006, we had decided to leave Guatemala and return, finally, to Australia. I wasn't sure what we were coming back to. All I knew was that I wanted to keep studying and write a PhD about human rights. The Fair Food scholar-activist had not quite come fully into view, but he was fast approaching.

GLOBAL CONSCIOUSNESS: LOCAL CONNECTIONS

Coming back to Australia in late 2006 brought a sense of déjà vu. Not because we were back in the land of our birth, but because we had to start all over again, rebuilding lives, careers, networks. Now with young children, the responsibility felt heavier, the challenge greater. This time we chose to go to a part of the country we'd never been, pulled there by a family connection: Bellingen, a small town of 3000 about 15 kilometres inland and 35 kilometres south of Coffs Harbour, on the idyllic New South Wales mid-north coast.

It was 2007. I had begun my PhD, fumbling for clarity and focus. Cracks were appearing in the stability of the global financial system, as the US sub-prime mortgages were beginning to turn sour and credit markets were seizing up. Our own main reason for living in Bellingen evaporated, as the family connection imploded in an angry series of recriminations. Old certainties were disappearing, in my own life and externally.

I needed a social network, and went along to some meetings of the local Climate Change Action Group, one of many established in the wake of Al Gore's *Inconvenient Truth* tour the year before. By this time I'd started reading about Food Sovereignty. I knew that it was a major global movement, and I knew that its head office was located (at that time) in Jakarta. I knew that no Australian organisation was a member of La Via Campesina, and wondered why. I had a strong feeling that my thesis research was going to be about Food Sovereignty.

At the Climate Change Action Group, they were talking about a campaign to make Bellingen plastic-bag-free. I wanted to talk about food and community gardens. They told me that a community garden had been attempted in Bellingen a few years before, but it had failed because of personality conflicts. I wasn't deterred, and began talking with some others about starting a local food network. We had a first meeting and more than 20 people showed up. There was a lot of excitement. One of those attending was the owner of the local wholefoods stores. He suggested we get hold of some food films and organise a local food film festival. Which we did, filling the Bellingen Memorial Hall over three Sundays in January and February 2008, culminating in a local food feast that served 140 people.

That film festival raised the profile of local food in Bellingen and beyond. I was invited to speak at a meeting of COGS – Coffs Organic Growers Society – and listening that night was the sustainability officer at Coffs Harbour City Council. She invited me and a number of others to work with Bellingen and Coffs Councils on an application to the New South Wales Environmental Trust. At that time the trust was funding a number of food-related urban sustainability projects across New South Wales. Our application was successful, and so the Coffs Coast Local Food Futures Alliance was born. Part of that project involved bringing speakers from other parts of Australia to the region to talk about their work, and that's how I met

Robert Pekin and Kirsten Larsen, two of the authors in this book, as well as Russ Grayson (Permaculture Australia and Australian City Farms and Community Gardens Network), who, along with his partner, Fiona Campbell, has been part of the journey of the Australian Food Sovereignty Alliance since the beginning.

I kept reading about Food Sovereignty, and was fortunate enough to be invited to attend a regional youth conference of La Via Campesina in Dili, Timor-Leste, in March 2009, where I met the then global coordinator, Henry Saragih, and other regional leaders. I went to Jakarta and interviewed some of their staff and visited some local farms. Just as I had been in Guatemala, I was deeply moved by the experiences of hardship and suffering in so many communities across south-east Asia, and equally by the passion and commitment of the farmer and youth leaders for positive change.

That trip confirmed for me that not only was my research going to be about Food Sovereignty, but the next part of my life – maybe the rest of my life – was going to be dedicated to sharing this story of struggle and hope and justice to as wide an audience as possible. People in Bellingen said, 'Why are you talking about Food Sovereignty? Why don't you talk about locavores instead?' But I was clear. Food Sovereignty was a global movement involving hundreds of millions of people. Australia was not part of it, and we needed to be part of it. The challenges we're facing, not just in the food system but especially there, have so much to do with excessive concentrations of economic and political power, and that power is more global than it is national. Those of us who do not have our hands on those levers of power need to understand that our strength, and our power, lies in our numbers. But we need to join together, to *connect,* to realise the potential of that power.

In those years – 2008 to 2011 – I discovered a few things about myself. That I had a gift for what they call in Guatemala *el poder de convocatoria,* which roughly translates as 'the power to bring people together'. That I could speak in public, develop a message and articulate it with clarity. That I could see the big picture, the whole, and also see where there were gaps, and where effective action might be taken. That I had a capacity to tell a story, and, indeed, that my own story might one day be worth telling.

And as well as realising some important strengths, I also found many things I needed to work on. That I was argumentative and at times struggled to see different perspectives. That I could exhaust people with my passion and voluminous writing. That I struggled to find balance in my life, and make sufficient time for my own family, and growing children. That as much as I thought I knew, in so many ways and on so many topics I knew so little.

AUSTRALIAN FOOD SOVEREIGNTY ALLIANCE (AFSA)

Essentially, AFSA formed, as many social movements do, in reaction to a proposal from government and/or corporations. Ours was the Gillard government's National Food Plan (NFP) process, which was launched after the August 2010 election. AFSA's 'founding document' was an August 2010 open letter we wrote to then Minister for Agriculture Tony Burke as the NFP process was getting underway. In that letter, we welcomed the government's intention to develop, for the first time, a whole-of-system food policy, but we said it needed to be genuinely participatory, and include the perspectives of the hundreds of thousands of Australians for whom food and farming were not simply about commodity production and 'free trade'.

We made submissions to the 2011 Issues Paper, and we attended round table consultations in 2011 and 2012. But by the time the Green Paper was due to be released, in July 2012, it had become clear that this consultation process was largely window-dressing. Far from being a whole-of-country conversation about the future of our food and farming system, as we had proposed, the NFP was very much 'business-as-usual'. The main driver was that 'the Asian middle class is getting richer', and therefore the main focus was on how Australia could increase levels of agricultural productivity, and sign more free-trade deals, to take advantage of this 'dining boom'. Food and farming were viewed, as they had been for decades, simply through the lens of 'commodity'. According to this perspective, we relate to the world around us, to our fellow humans, to ourselves even, primarily through that mythical and magical substance known as 'money'.

Earlier in 2012, a Canadian by the name of Amanda Sheedy had visited Australia and talked to some people about a process she'd helped coordinate in Canada called the People's Food Policy Project (PFPP). This involved 350 'kitchen tables' of groups of ten people – friends and neighbours – discussing their concerns and priorities about Canada's food system, according to the principles of Food Sovereignty. The PFPP was supported by an indigenous circle of elders, and took direct advice from farmer reference groups. Teams of volunteer academics and others helped synthesise and write up all the notes generated from those kitchen tables into ten detailed policy position papers, and a single People's Food Policy for Canada.

I didn't get to meet Amanda, unfortunately, but I was excited when I learnt about the PFPP. My thoughts were, 'If they did that in Canada, why can't we do that here?' In fact, that's become something of an operating principle of mine, as I am now working with colleagues on a campaign for a Local Food Act and Fund in Victoria, inspired by the precedent set by the Ontario government in 2013.

We didn't have a $250,000 grant to fund a PFPP in Australia, but we didn't let that stop us. We had a national network of community gardeners, Transition Town groups, small farmers, health professionals, food social enterprises, and students and academics. We thought, 'Let's give this a go and see where it leads.' So we did: four of us worked on a *Draft Discussion Paper of Values, Principles and Best Practice,* which we circulated through our networks with an invitation to hold public forums to discuss the issues. SBS TV's 'Garden Guru' Costa Georgiadis wrote a beautiful foreword for the document. Forty public forums were held and 600 people participated from September to December 2012. Then Carol Richards and I spent days during December 2012 and January 2013 writing it all up into a *People's Food Plan Working Paper,* which Michael Croft launched in the Sydney Town Hall during a visit to Australia by 'heretic' US farmer Joel Salatin, in February 2013.

The People's Food Plan process generated a lot of excitement in lots of places. This was something new. It was Australia's first 'crowdsourced' food policy document. But what to do next? A 'movement' has to keep moving. A small group of us decided to self-fund a weekend strategic retreat in Bateman's Bay, at the beginning of February 2013. Almost coincidentally, shortly before that retreat, a couple of us had got hold of a wonderful little book called *Reimagining Change: Narrative-Based Strategies for Social Change.* Its central premise was that if you want to work for large-scale social change – transformation – you have to engage at the level of language: narrative, metaphor, myth, meme and discourse. You have to critique and deconstruct the 'control myths' that sustain a dominant system of power, and create your own myths about the new system, the fair system, that you want to see emerge and flourish. Humans are creatures of meaning; we live our lives according to the systems of meaning that, typically, we inherit and don't question, and those meaning systems are created and sustained through language. Words are immensely powerful. Stories are powerful, because, if effectively told, they reach people at the emotional, intuitive level. And that's when people can be motivated to get involved and work for change.

We workshopped AFSA around the resources in that book for the 40-odd hours we spent together that weekend. For me it was a like a light being switched on in my head. *This* was what we were doing. *This* was my role, the contribution I could make. We were – are – creating new concepts, new images, in the Australian conversation about food and farming systems. We searched for a *meme* – a cultural container of meaning: something that would resonate in the Australian cultural context, that would not require detailed explanation. Food Sovereignty was not that meme; it was too new, too unfamiliar. But Fair Food could be, because notions of 'fair' run deeply through our cultural psyche. Most of us have a notion of what is 'fair' and 'unfair': we don't like seeing people ripped off, and we have an image of ourselves as a reasonably egalitarian society, even if we're far less egalitarian now than we were, say, a generation or two ago.

So out of that weekend and in subsequent discussions with colleagues involved in the People's Food Plan process, Fair Food Week emerged: a week in the calendar year during which we could throw the spotlight on women and men around the country who were already making changes in their communities to support the emergence of a Fair Food system. Our tireless web volunteer, Fiona Campbell, created a self-organising website (www.fairfoodweek.org.au), on which groups could register and upload their events. In the first year, 19–25 August 2013, 112 events were hosted in every state and territory, with an estimated 15,000 participants. These events included tours of community gardens, film nights, speaker forums, farm dinners and workshops, and dedicated farmers' markets. The next year, 10–19 October 2014, we had 100 events, with 25,000 participating.

We had created a space in the calendar year to throw the spotlight on Fair Food in Australia. People were now talking in terms of Fair Food, and what that meant. I participated in a couple of political candidates' forums during the Victorian state election of November 2014, in Carlton and Eltham. The topic was 'Fair Food', and the candidates were asked to explain how the policies of their parties were contributing to the building of a Fair Food system in Victoria. Of course, you might say, talk is cheap, and they were speaking to particularly engaged audiences on each occasion. But for me, the fact they were using *our linguistic and mental concepts* was highly significant. Through the exercise of our cumulative creative powers and efforts, the new discourse of Fair Food was starting to take a hold in the culture. And, if it spreads far enough and wide enough, that's what brings about a shift in our collective worldview. A paradigm shift. It all starts, as Charlie Massy (whom you'll meet in this book) says, with that 'one square foot of real estate between our ears'.

Now I'm going to let my friends and colleagues who are walking this journey with me – who in a very real sense are creating the path by walking it – take over the narrative. I'll introduce each one and tell you how their story fits with mine. Come with me and meet them!

PUTTING THE 'CULTURE' BACK INTO AGRICULTURE

ROBERT PEKIN WITH EMMA-KATE ROSE

The year 2009 was a seminal year in the progression towards the 'new normal' of extreme weather patterns and events, with a record six floods on the New South Wales mid-north coast and the tragic Victorian bushfires. It was also the year I connected with La Via Campesina, and resolved to bring Food Sovereignty into the Australian public and political discourse. So it was no coincidence that it should be the year I met Robert Pekin, one of Australia's Fair Food pioneers: a man who has dedicated the greater part of his adult life to bringing to Australia a model of food distribution that truly connects farmers with city and townsfolk.

The Coffs Coast Local Food Futures Alliance had a budget to bring speakers to the region. I'd heard about Food Connect Brisbane, and I asked Robert to make the six-hour trip down to talk about their social enterprise. I remember the excitement among the 40 people who gathered at the Thora General Store as Robert told his story about the risk-sharing model of community-supported agriculture, and then sketched out on butcher's paper what the local producers might aggregate as a collaborative business selling vegie boxes and other produce direct to the local community.

Not long after, Robert invited me to go to Brisbane to meet his wonderful partner, Emma-Kate, and their young family, and learn more about Food Connect. It was the first of many trips I would make to spend days and evenings with Robert and Emma-Kate, talking about Fair Food systems and models over delicious meals in their West End kitchen.

Robert was with me and Russ Grayson in Newtown in July 2010 when we decided to send our open letter to Tony Burke, which resulted in AFSA. In February 2015, Food Connect Brisbane celebrated its tenth birthday, a remarkable achievement in light of its beginnings and the very challenging business environment in which it has always operated.

In this chapter, Robert and Emma-Kate generously share a great deal of what is an extraordinary story of loss, pain, and hardship, but also of the joys of perseverance and commitment through adversity, holding fast to ideals, principles and a vision of the fairer food system this country so desperately needs.

***

> *You never change things by fighting the existing reality. To change something, build a new model that makes the existing model obsolete.*
>
> Bucky Fuller

BORN A DAIRY FARMER

Being born into a fourth-generation farming family in Victoria's Western Districts was not the easiest upbringing. The land was originally owned by pioneer pastoralist William Robertson, who divided the estate among his four sons. Our part of the 'estate' was originally known as the 'The Hill', and was bought by my great-grandfather Sandy (Alexander) Pekin, after James Robertson's death. Sandy's son – my grandfather Daniel Pekin – was gifted the property when he married in 1919, and he commenced dairying as the principal farming activity.

My parents were traditional, hardworking types, running the household according to strict Catholic doctrine, and through that lens, they built a tough but relatively idyllic life on the land. We milked cows and grew crops, and in my younger days there was a great sense of community, cooperation and connection with neighbouring farmers, the cooperatives and the local townspeople. It was the 1960s. These were the days when the local AFL footy and netball clubs were fully subscribed, and the country tennis circuit flourished.

I was the eldest child of nine, and, at 16, desperate to see a bit of the world (and escape the priesthood), I joined the navy, spending over a decade training and working as an aircraft mechanical engineer. It was here I met my wife, and we decided to begin our life together in central Queensland where she was born. I had just started my own marine repair business, and in 1992 my mother called with news that they were taking on a share farmer. None of my siblings were interested in taking on the farm, and after mentioning it to my wife, she became interested. She knew I was struggling in the central Queensland heat, and, after a few days of discussion, we agreed with my parents to return to my childhood home to start farming.

RETURN OF THE PRODIGAL SON

I was excited to be back on the land of my birth with a chance to create my own destiny. I got involved in everything to do with re-educating myself on dairy farming with gusto. However, I was soon starting to question modern farming practices and, after about nine months, realised that the chemical industry had an all-pervasive presence on our farm – even the field days were hosted by chemical companies.

When you grow up on a farm, your dad's running the place, and you're doing all the farm chores without question. After 12 years away, I had gathered a bit of confidence and was asking a lot of questions. What shocked me was the way I was treated by the bigger farmers and the representatives from the chemical companies for being slightly curious. At the same time, my father's pride in the 'prodigal son' returning to the district was beginning to show cracks. I must have been a right pain in the arse, as he was defaulting to treating me as the 16-year-old who'd left the farm. The build-up of family tension combined with the general stresses of farming life soon forced my marriage to fall apart. After a period of sadness and internal questioning, I resolved to continue share-farming with my parents.

It was during one particular field day, when I questioned the use of nitrogen, that I got a true insight into the attitude of the chemical industry. I'd just read a journal article explaining how nitrogen 'freezes' the soil and the microbial life within it. The chemical company rep, who was also our dairy company's field officer, turned to me aggressively and said, 'Well, the soil is just a medium to grow food on, and in 10 or 20 years or so, when it's all buggered, we'll just move onto other land.' He could justify this stance, because at the time there was a lot of cheap country with shallow underground water being bought up by dairy farmers in south-eastern South Australia. With high milk prices at the time, it was a bonanza for both the chemical companies and stock and station agents.

Not long after this exchange, a friend of mine told me about his father dying of cancer, and how his mother swore it was from the chemicals. His family wasn't alone in the district. Eventually, a group of 15 to 20 local farmers got together to organise alternative field days, and we formed the first Australian Sustainable Farm Managers Association, which conducted the first organic dairy farmers course. My journey into sustainable agriculture had officially begun, and there was no looking back.

I attended David Holmgren's Permaculture Design Certificate course in 1995, and I was now seeing the farm through my cows' eyes, and understanding the value of looking after the soil. I began repairing the landscape through my membership in the local Landcare group. I read everything I could get my hands on, and after a while got invited to host my own radio show for three months called 'Milking Time', provoking new (and old) ideas on dairy farming. I'd invite alternative farmers, town vets and agricultural field officers to have conversations about their observations and practices, and it started a bit of a groundswell. Although seen as controversial at the time, all we wanted to do was to make people question the conventional wisdom and think a bit deeper.

I began to understand the distinction between chemical farming and natural farming. My general observation was that using chemicals or taking advice from multinational companies forces you to become a specialist farmer, whereas going organic requires you to listen, observe your land holistically, and understand the life cycles of all the critters, such as the red-legged earth mite. It's an endless field of study to become an ecological farmer.

After continual conflict with my parents over my new vision for the farm, our working relationship was starting to break down. My parents basically gave me an ultimatum: farm it according to tradition and 'conventional wisdom', or buy the farm outright. I took the latter offer, which also meant buying each of my eight other siblings' share of their inheritance. This was not an affordable option; however, I stubbornly and optimistically forged ahead, paying over and above the initial valuation provided by the family farm-transfer specialist. I was now in debt to the tune of about $1.2 million.

CUSTODIANSHIP

I was excited, and I could now throw myself into trying out new things. I had a great supportive group of fellow farmers, my siblings were happy with the arrangement and I enjoyed the company of my youngest brother, Anthony, working alongside me. Added to that, decent autumn rains had fallen, there was good pre-winter grass cover and milk prices were at a five-year high. The least of my concerns was a million-dollar undertaking.

The 'whole farm plan' I'd been working on – everything I thought agriculture should stand for – I could now put into action. I was already on a roll, planting out wide shelterbelts and agro-forestry projects, hosting WWOOFers and other city folk keen to help out, and I began recycling projects of wash-down water from the yards. I spent the evenings reading all subjects, from Henry Thoreau through to PA Yeomans. I trialled new natural fertiliser regimes, minimal till cropping, and under-sowed pasture seeds into summer crops, so I would not have to cultivate again in the autumn. An ambitious attempt to change the calving pattern – from autumn to spring, to autumn only – was going to plan.

However, my enthusiasm was short-lived. Two months after I purchased the farm, the opening milk price was 20 per cent lower than the previous year. A long, wet winter followed by a snap heat-wave in early spring brought on the first drought. After another two years of drought and a further 15 per cent fall in milk prices, I was pushed to the brink. Hay and grain prices tripled, and despite my paying back over $370,000 off the purchase price to my parents, our relationship was still acrimonious. They had given me some financial relief in the second year of drought but after the third year of tough conditions were not prepared to look at solutions.

Eventually, I had to lay off my two dedicated farm workers, my good mates Dominic and Nicky. I had to force my cows to eat rice straw, and milk my herd of 310 cows on my own twice a day, all the time hoping for a change. Like the eternally optimistic farmer you often see in Australian literature and TV, I was determined not to give up. It was February 1998 – the month of my lowest cash flow – when I received a letter from the bank's headquarters in

Melbourne. It demanded a full payback of the loan and my overdraft over the following three months. Most of the herd were getting ready to calve, so they were in between milking, and, according to the long-range weather forecasts, I was looking down the barrel of another dry autumn.

I was devastated, but still determined. I sold 100 of my best cows to pay back some of the money, and this provided temporary relief. Slogging it out on my own caused me to question a lot of things, and the isolation only made my state of mind worse. It was then I really understood that I had no connection with the people who were drinking my milk. The closest I ever got to them was waving the milk tank driver off as he passed through the farm gate. At the same time, the CEO of our demutualised cooperative was on a million-dollar salary. I felt there was little respect for my efforts and questioned whether I was of any use to society. I was hearing of suicides in the district. One was a footballer I had played against over the years. I felt disconsolate and invisible, and for a few months suicide was not far from my thoughts. It was a roller-coaster of lethargy, rage and desperation.

The American philosopher Ralph Waldo Emerson once said, 'When it is dark enough, you can see the stars.' It describes the moment I had hit my lowest point. The banks were threatening to foreclose on me, my parents were jostling to take ownership back off me, and I was about to do that unforgivable thing: I was 'that' generation to lose the farm. I'd withdrawn from family and friends and lapsed into a haze, milking cows on autopilot.

One night, I decided to do away with myself and was heading over to borrow my neighbour's gun, when I began to have second thoughts. Halfway there I stopped and returned home. I stepped inside and went back out again, this time to the top of the old pea paddock. Lying on my back, I tried to comprehend everything. I looked up into that immense, endless sky and suddenly realised how pointless it would be to sacrifice my life. I saw the stars.

The next day, I met with my solicitor and said, 'That's it. I'm out.' My solicitor protested initially, saying I could pursue legal action against my parents, but I just wanted it over. It took a couple of weeks to tie up loose ends. Eventually, I was towed off the farm in an old Bedford bus that had been used to house volunteers. I had nothing but my backpack and a huge debt.

INTO THE WILDERNESS

I was an angry man. I'd severed ties with my family, resolving to spit on my father's grave, and was on the run from creditors. Even though suicide wasn't an option anymore, I was very lost. I drifted around, accepting the hospitality of good friends. My brother Tim took me into his home and allowed me time to get through my sadness. It was during a conversation with my uncle Noel that he could see I needed a lot of solitude, away from the everyday routine of society. He told me to clear off to Tasmania for a while. It sounded like a good idea.

Not long after, I arrived in Devonport with all my belongings in a backpack, a quarter of it filled with books. With R Buckminster Fuller's *Critical Path,* EF Schumacher's *Small Is Beautiful* and many other titles, I looked for answers. Not long after setting out, I met an Indigenous man and learnt about their ways: how to ask the mountains for permission to climb them, how to listen to the signs in the bush all around me. I noticed that Indigenous people don't look down when they walk. They just let their feet fall and magically curl around whatever they step on.

As I journeyed deeper into the wild landscape around me, I grew a second skin. I started out sleeping in a tent, then I moved to being outside in my sleeping bag. After a while I could fall asleep anywhere. I was in tune with the bush animals, the Tassie devils, goannas. Even the biting insects that perpetually surrounded me didn't bother me. I was at one with the ecology, passing through it, noticing every detail and hearing the rich life of the forests. I began to feel deeply dispossessed of my land back in Victoria. It dawned on me that what I was feeling was just a fraction of the enormity of the experience for Australia's first peoples' systematic dispossession for over 200 years. Being deeply present with the land in Tasmania allowed me to understand fully how interconnected we really are with our environment.

When I left the farm in Victoria, I was not in a fit state to finalise the financial situation. Walking away, in the end, actually saved me from declaring bankruptcy – which would've been even more of a disgrace in my book. Through all the walking and meditation I began to lose my anger. I was now aware that I needed to stop blaming and take responsibility for the mess I had gotten myself into. This also meant taking responsibility for being pig-headed when buying the farm off my father, and the high price I agreed to pay my parents for the farm. They had all been my decisions. After these epiphanies I was able to go back and sort out a few things with my parents and others. I called one bank manager and said I would pay everything I owed. It would take time, but everyone would get paid. My parents had already settled with the other bank.

When I returned to Tasmania, I tried my hand at a few interesting jobs, mainly in farming and the seafood industry. However, a turning point was meeting a German backpacker who mentioned an interesting project at a farm south of Hobart. Not long after, I found this small farm on the grounds of a local Steiner school. The farmer was getting a fair price and members of the community would help out, as well as buy a weekly share of the harvest. The hairs on the back of my neck instantly stood up. This was one of the first organised community-supported agriculture projects in Australia. Right then and there, I decided this was what I wanted to do.

COMMUNITY-SUPPORTED AGRICULTURE (CSA)

I quickly learnt that CSA is a model where a local farmer and supporters in rural and urban communities get together in the spirit of mutual support and risk-sharing. It is a radical attempt at placing control of the food system into the hands of people. Farmers carry the burden of risk in the farm-to-table chain, and this is a way for people to take responsibility in a positive way. While there are many versions and forms of CSA, the true, basic definition is a partnership between the farm and consumers where the risks and rewards of farming are shared. I did my research and found that the globalised, industrial model of our food system was responsible for many inequities. The pieces in the puzzle of my own story in the dairy industry started to fall into place, and I felt a renewed sense of purpose. I learnt that CSAs originated in Japan in the mid 1960s. *Teikei,* as the model is known there, means 'cooperation' and is now reaching over 2 million people in Japan. Switzerland was the next country to initiate it, and from there it spread throughout the United Kingdom and the United States via the biodynamic movement, originally founded by Rudolf Steiner. The United States has so far achieved the fastest growth in the number of CSA initiatives, although it still only accounts for 1–2 per cent of the market. Across Europe, the estimated number of CSAs is around 4000 farms and 400,000 consumers.

After experiencing the work at the school's farm, led by Mark Patton and his partner, Kathy, I started looking for land in the area and knocked on a few doors. Eventually, a farmer named John Maddock just south of Kingston opened his door to me, and I explained what I was planning to do. I said I'd run the paddock as a CSA initiative, and he said, 'What is that? Some kind of communist thing?' And I said, 'No, I don't think it's that. It's just a new model of farming. It's from the philosophy of social renewal. It's a risk-sharing opportunity for subscribers and farmers.' John's eyes lit up, and he let me have a six-acre paddock over the back for six months, rent-free, with free electricity and water, as long as I worked a day a week for him welding and repairing machinery. It was a great deal, and I felt lucky that he was open enough to give me a start.

My recovery was initially facilitated by my time spent alone in the Tasmanian wilderness, but this small-scale, single-farmer, CSA initiative on that piece of leased land near Kingston was the catalyst for propelling me forward. I used Groh and McFadden's book *Farms of Tomorrow Revisited* as my new bible. Working with the soil again healed me, and the relationship I formed with John, my landlord farmer, created a supportive working arrangement for the idea to take off.

After initially cultivating the land with John's equipment, I hand-formed all the beds for the market garden. I funded materials and infrastructure by working part-time for a local seafood company. I started growing salad vegetables and leafy greens, selling them on the side of the road, and to a couple of local cafes. With the help of a motivated couple I'd met from the school farm, Bob and Joy, we spread the word looking for local subscribers. Together we grew 70 different types of vegetables. Within those types, we'd have another 7 to 8 varieties of carrots, 12 varieties of tomatoes and 20 varieties of garlic. We experimented with varieties to see what would grow well on the farm and be most productive. We were even successful in growing some watermelons – very small watermelons – in Tasmania! With the help of a couple of older neighbouring market gardeners, I also learnt all the tricks to harvesting and bunching. I listened with interest to their stories and the way their lives were impacted by the declining terms of trade from the industrial food system.

Community-supported agriculture: principles

According to the first international CSA symposium in 2011, the different CSA movements recognise the following four fundamental principles:

1. partnership: usually formalised as an individual contract between each consumer and the producer, and characterised by a mutual commitment to supply

one another (with money and food) over an extended period of time

2. local: part of an active approach to relocalising the economy. But local in the CSA movement is not restricted to a geographical meaning. The idea is that local producers should be well integrated into their surrounding areas: their work should benefit the communities that support them

3. solidarity: between producers and support groups and involving:

3.1. sharing both the risks and the benefits of a healthy production that is adapted to the natural rhythm of the seasons and is respectful of the environment, natural and cultural heritage and health

3.2. paying a sufficient fair price upfront to enable farmers and their families to maintain their farms and live in a dignified manner

4. producer–consumer tandem: direct person-to-person contact and trust, with no intermediaries or hierarchy. It's the complete opposite of the faceless exchanges we experience under the centralised market and monopsonist system we currently have in Australia.

After a year or so of operations, our new CSA was beginning to do well enough to think about the future, and I started to talk about bigger plans. It was not viable enough for all of us to make a living, and I felt we needed to get a bit more serious about expanding the operation. The ideas I had didn't go down as well as I'd planned. I took a couple of weeks off to celebrate my brother's wedding back in Victoria, and returned with the team in conflict over my proposed vision. In the end, the future could only be decided by our landlord, John. My vision included opening a dairy, creating a fully closed loop system on the farm, building a community space and eventually buying some of the property. The alternative plan was to continue the current model as it was, and in the end it was John who reluctantly decided that my vision was too big or unrealistic to give it the go-ahead. The team paid me out with my share, but I was heartbroken again. It hit me hard.

AGRICULTURAL JOURNEYMAN

It was the year 2001. I packed up my old Dodge ute with all my belongings and headed back to the mainland where I continued as an agricultural journeyman. I was homeless and farmless (not quite harmless) again and desperately keen to keep this dream alive. By that stage word had got out about the model, and I travelled all over Australia advising farmers, giving talks and helping them set up their own CSAs. In fact, I think it was the only thing I talked about.

During that time, the thought came to me that I was nearly 40 and had worked physically hard all my life, so I resolved to take some time out to reflect and think about how I could both reach out to the thinkers of the world *and* apply my preferred pragmatic, 'doing' style. I continued my studies in the philosophies of Rudolf Steiner, enrolling in the foundation year of Anthroposophy, and, as circumstances had it, ended up in Brisbane for the birth of my first child, Dougall.

Making a new life in Brisbane was difficult at first. Carrying the hurt from my attempts at new ways of farming, and disappointed with limited access to my son, I resolved to keep pursuing my purpose. I was consulting with farmers in Beerwah on the Sunshine Coast, and being asked to speak about CSAs at a couple of events, when I received a call from Kristen Lyons from Friends of the Earth.

Friends of the Earth had just launched its publication *Towards a Community-Supported Agriculture* as part of its Food Justice campaign, and Kristen invited me to get involved. While I'd been to a few protests in Melbourne over the years, it was my first real interaction with city people, academics and activists. I was enthused by the attitude and response of these city people to the idea of CSAs and we set about creating a plan for the future based in Brisbane.

The question for CSA farmers is not only, 'How do I get one started?' but also, 'How do I keep it going?' The amount of work and the level of responsibility to your CSA members can be tremendous. At this point, I was keen to visit many of the CSA farms in Australia to see how they were travelling. Farm after farm, I was disappointed to find that they had struck limitations and struggled to remain viable. It appeared that despite the general acceptance of the model and the good intentions of their supporters, two common challenges in the Australian context made it unviable for single-farmer CSAs.

First, the burden of growing a huge diversity of produce in sequence to meet consumer demand on a weekly basis was an almost impossible task for one farmer. Gradual monoculture farming has left huge gaps in knowledge and experience in the Australian farmers' skill set. An added complexity was that consumers in Australia seemed reluctant to ride out the lean times in between the seasons, and would drop off their subscriptions.

Second, many of the failed CSAs weren't located close enough to densely populated centres. The sheer distance between some farmers and their consumer shareholders was too great and reduced the opportunity to establish a strong community base necessary to support the farmer and the farm's needs. And so it occurred to me that we needed to tweak the model to better suit the Australian context, while staying true to the principles of CSA.

I envisioned a hybrid model that brought together a number of growers with an even larger pool of subscribers. The only way it could happen would be to set up an enterprise in a densely populated area that was as 'efficient and equitable' as possible in the chain. Acknowledging that most farmers 'just wanted to farm', this enterprise would be made up of a large 'core group' – a facilitator, or ethical 'middle man' – looking after the packing, marketing and distribution on behalf of and in consultation with the farmers.

THE BIRTH OF FOOD CONNECT

One morning in July 2004, ABC local radio announcer Steve Austin was interviewing the Queensland Fruit and Vegetable Growers' chief advocate, Mark Panitz. He was vehement about the federal government's backdown on enforcing a 'mandatory code of conduct' for all central markets around Australia. The markets lobbied hard, and it was getting very personal. I called the radio station to ask more about it, and not long after the producer invited me to speak to Steve the next day.

What was supposed to be a ten-minute slot ended up filling the rest of the hour. Immediately afterwards, the phones lit up the board with passionate people wanting to talk to me about getting something going in Brisbane.

I had already spoken to a group of farmers in the Lockyer Valley to help them set up a multi-farmer CSA venture, which would sell into Brisbane, and now with the support of a passionate group of city mums, as well as some other key farmers in the region, we commenced setting up Eco-Foods as a joint venture. Unfortunately, there was already internal conflict among the Lockyer Valley farmers, with the leading farmer nearing bankruptcy. Despite our best efforts to assist, the issues were too complex and, soon after, the supplier end of our new model fell apart!

From the ashes of this disaster, the emergence of the first iteration of Food Connect was born. The initial core of Brisbane-based subscribers were committed to the success of the model. They'd had a taste of it, and could see its potential. They insisted that I continue and they would assist with the distribution side of things. So I set about recruiting farmers from all over south-east Queensland, from contacts I had established over the years. At the same time, the Brisbane mums set up drop-off points across Brisbane, which we called City Cousins – because we all have 'Country Cousins', don't we? Amazingly, we seamlessly transitioned the deliveries from my one-bedroom flat and it expanded rapidly to 300 subscribers by October. Remember, this was long before Facebook, Twitter and Instagram were twinkles in geeks' eyes. Things rapidly got out of hand. We couldn't keep track of the cash coming in, and our systems were getting shaky. Then a miracle happened. A huge storm hit Brisbane and a lightning surge fried the laptop on which all of our customer and ordering information was stored – with no back-up! I had been working crazy hours picking up produce from farmers at 2am, working through the day and delivering orders to City Cousins, and I had pushed myself to the point of physical and mental exhaustion. We were all struggling to keep up demand with only a few very tired part-time staff; we enforced a break to regather and plan for the 'relaunch'. Once we stopped, I think I slept for three weeks!

RELAUNCH

After such a short time in operation, we had fostered trust and commitment in complete strangers across Brisbane and in our new group of farmers across south-east Queensland. We held a number of founding meetings with some business advisors, IT specialists and other supporters where we planned the structure of the organisation and physical infrastructure requirements, as well as testing the financial model. Out of these meetings, the business advisory group was formed, and I was given the responsibility of running the enterprise.

A business plan was drawn up, and a trip to Bendigo Bank booked to ask for start-up capital. In the meeting with the bank's branch manager I was asked who our demographic was. I naively answered, 'Anyone with a mouth!' Of course, we were knocked back, and so with $5000 worth of 'delayed subscriptions' (or seed loans) from the advisory group and a few committed supporters, Food Connect Pty Ltd was born. We kicked off on 28 May 2005.

We shared some warehouse space with local co-op Reverse Garbage in Salisbury, and I secretly lived there for a year while building up the business. I would take deliveries from small farmers at 3am; however, for the bigger deliveries, the layout of our complex couldn't accommodate larger trucks. So I set up an elaborate scheme with the farmers and independent transport companies to stay under the radar of the Brisbane market mafia by hiding pallets inside cold rooms, in return for cartons of beer once a month.

To receive produce in the central markets, the normal practice is to register with the market management, pay a licence fee, and buy through one of the market agents. I was set on minimising the 'sets of hands' our farmers' food had to pass through, so I had to break some rules.

It's a dog-eat-dog environment, a market gone mad, with no sense of cooperation. So it was fascinating seeing this barter economy start up. The truckies and forkies would let me know of farmers they knew who might fit the bill for us. They could see that we were growing, and they helped us, picking up directly from farmers like they used to do in the old days. So these relationships were coming back, which I'd never anticipated when I started out. It was another part of the local economy that had been completely done over by the system, and it was a vital part of the link. Sometimes, I'd play the game and buy produce from market agents purely in order to get the farmer's details on the box, and then I'd go about paying them a visit.

I knew it couldn't last like this, and we'd grown enough by now, so we did end up paying a registration fee to collect our produce via the Brisbane Market Unloading Service. Once we moved into premises large enough to accommodate larger trucks, we started receiving direct deliveries from our farmers and independent transport companies. The drivers and farmers now let themselves in and out of the warehouse at all times over the weekend and through the middle of the night to make deliveries.

MAKING THE EXPERIMENT REAL

For the first few years we experienced year on year growth rates of 50–80 per cent. We were the only enterprise in Brisbane with the unique offering of locally sourced, naturally farmed boxes of fresh produce, using our distribution system of City Cousins. In line with CSA principles, we focused predominantly on retail sales to subscribers who paid upfront. At the time, the only two other box schemes in Brisbane offered full choice and home delivery. We went about working with the law firm Mallesons who helped us design our constitution, pro bono. We wrote into the constitution a number of provisions to guide our social purpose: an asset lock; a 2:1 wage ratio for everyone, including me; a circle of elders/advisory board; and all profits directed back into furthering the goals of the business.

While this new way of doing food was exciting, we knew we were creating something that offered a transition model, given the scientific reports coming out on climate change and peak oil. It embraced many of the principles of Permaculture – costs were kept low by sourcing second-hand equipment. 'Fix 'n' mend' is a constant mantra at Food Connect, along with 'no waste', 'making hay while the sun shines', 'reusing, recycling' and 'valuing diversity' – both with our products and our people – and using small and slow solutions.

It wasn't an easy ask for consumers, but enough city folk understood it and embraced it. To lower the barriers for entry, we set the upfront subscription as a minimum of four weeks; however, many customers opted for longer subscriptions, sometimes buying a year in advance. It was a great model for bootstrapping the enterprise's growth.

HIGHLY NETWORKED, DISTRIBUTED AND LOCAL

Food Connect works in the vein of the *Small Is Beautiful* philosophy or what others like to call 'economies of community' rather than the old model of 'economies of scale'. What we've seen and experienced over the last 50 years or so is the slow death of small industries serving their regions. Towns all over Australia have lost their local bakeries, abattoirs, creameries, butter factories, grain mills and greengrocers. Our culture is the poorer for it – everything has a 'homogenised' look, feel and taste to it. The implication is that we are also homogenising our brains.

Economies of community

To democratise the food system, we need three things:

1. a network of ecological farms in our bioregion

• growers get paid fairly and invest more in their local community

• growers have more control and less dependence on the central market

• growers are acknowledged, form relationships with consumers and feel belonging and an integral part of their community

2. highly local distribution channels: many growers to many distributors builds resilience into the supply and demand of the system

3. motivated consumers who:

• value locally grown food

- maintain quality control through direct feedback to growers, thereby maintaining high standards of ecological practice and growing skills
- support small, locally owned distributors
- recruit others to join the movement.

I knew that distribution was where most of the inequities were in the food system, and localising a food system would address many of them. I figured if Food Connect could demonstrate the 'how', then we would have achieved a lot in terms of rebuilding the lost infrastructure – both physical and social – that once served our communities so well.

In 2007, we were given notice to leave the warehouse we shared with Reverse Garbage, and found an ideal place not far away. It was important for me to know the landlord by name and establish a strong relationship. We struck gold in Wayne, who shared the same passions as I did for doing business differently, particularly Ricardo Semler's work, encouraging a democratic workplace, empowering employees to come up with solutions on the go, and taking responsibility for their roles. Unfortunately, shortly after signing a five-year lease, we found out we had to pay three months' rent in addition to the bond, due in two weeks!

We had no money, but we liked the feel of the warehouse and the landlord, and really wanted to make it work. So once again, we went to a couple of banks who knocked us back, and then we asked Social Ventures Australia, who were asking me to get involved with their projects at the time, but we were told that they wouldn't be able to raise the money in such a short period. The only thing left for us to do was to go out to our community. The response was incredible and within two weeks we'd raised nearly $30,000, which we paid back over time in subscriptions or as loans.

LOVE WORK AND OUR CITY COUSINS

I often say that Food Connect work is love made visible, and I mean that in terms of love on every level. It's not something we farmers tend to draw on to describe what we do. It's in the doing of our work that we can demonstrate the love we have for the farmers that grow our food, and that in turn influences the way they tend the soil, care for their plants and animals, which, through the process of eating, influences our collective love for the planet, love for our children and love for our whole community.

The City Cousin network initially came about as a practical way to deliver food from the farmers to consumers. However, we quickly realised that it had a far deeper role in connecting neighbours within the city – what we call 'lowering the fences'. For decades we've witnessed the replacement of the white picket fence (or no fence at all) initially with chest-high walls with vertical or horizontal gaps and, later, higher walls that completely shut people in (and others out) to the point where the design of our suburbs is preventing people from knowing their neighbours.

I could see that City Cousins could reverse that trend and, despite it being attacked as a business weakness (by consultants and other 'experts'), we were pretty confident that it would have a positive impact, in terms of the core principles of CSA – sharing the risk and building community.

It has not been easy. Love never is. We've had all types join the City Cousin network, some risking the integrity of the business, but many more inspiring individuals who have initiated amazing projects. From school-based, adopt-a-farmer programs to community relief programs, neighbourhood cook-ups, organising protests and food justice campaigns, hosting Transition Town meetings, forming cooperatives and bulk buyer groups, underground supper clubs, backyard harvest swaps and yarn-bombing, right down to lots of sharing around how to feed a newborn and trustworthy babysitting contacts. This network has fostered and complemented the broader, grassroots social environmental/food justice movement in Brisbane.

For me personally, there was an added benefit. My meeting with one particular woman opened the door to a new love of the romantic kind. It was 2007 and Emma-Kate Rose had just launched her new car-sharing business and was running the Ashgrove

Climate Action Group when we met at a forum she helped organise at the Griffith Eco-Centre. Emma-Kate had never heard of Food Connect, but once we got talking she immediately expressed enthusiasm for becoming a City Cousin. I had some ideas of my own. After my failed marriage I spent many years believing I was too busy or too unsuitable for another long-term relationship. In the intervening years, I had made a few attempts at love, but my record was not great. Even though I knew I was under the spell of 'once bitten twice shy' (and all the other boring excuses), I felt a new determination to not let this opportunity go by.

My first 'visit' to Emma-Kate's place was under the pretence of 'supervising' our City Cousin coordinator Ian Dalby's inspection and induction. I fell in love after about 15 minutes, and, as Ian put it, he may as well not have been in the room, such was the connection between us. Not long after, I mustered up enough courage to ask Emma-Kate out, and pretty soon we were inseparable. It was a fantastic for me to find someone who was not only beautiful and radiant but also passionately committed to a better world. From a practical perspective (not that I could think that clearly in the early days) I knew this 'someone special' needed to appreciate the long-term nature of the work I was doing, and Emma-Kate did that in spades. The rest, as they say, is history, and over time she has gradually become a big part of Food Connect's evolution, eventually taking on the role of General Happiness Manager. She's taken responsibility for professionalising and improving our processes and established a more sophisticated and interactive marketing presence.

Seven years on, and we're writing this story together. We've integrated our lives where our blended family of four children (one of our own) now lives at the Food Connect homestead. While it would be fair to label both of us workaholics, we don't subscribe to the idea of a work–life 'balance'. We believe it is all 'life' and, like farming, there are times where the harvest means you need to make hay while the sun shines, and take the time to rest, reflect and be together when it's cold and wet, or in Brisbane's case, hot and wet! In my view, separating work from life is a matter of discipline, and only time will tell whether we can succeed in a city-based business, establishing family and work rhythms around the seasons and the people we work and play with.

GROWING PAINS

Looking back now, it is clear that the early success of the enterprise was due to two factors. First, word of mouth was strong, complemented by my speaking schedule and conversations across the distributed network of our City Cousins. Second, the difference in quality and freshness in the produce was so apparent in the early days, and people got naturally excited about that. At the time, people were so conditioned to eating food from somewhere else that, unless they grew it themselves, they had little understanding of what 'fresh' actually tasted like.

After weathering the effects of the 2008 global financial crisis, the period from 2009 to 2012 was incredibly tumultuous. A rush of media exposure, speaking gigs and events saw our numbers grow rapidly. Soon we were at 45 staff and delivering over 1000 boxes a week, plus an increasing inventory of value-added products. We received a grant to build a commercial kitchen, but its delayed construction and our (over) commitment to employment outcomes blew out our costs. At the same time we had difficult governance issues. While trying to establish a governance structure based on democratic principles and a lateral organisational structure, I was being accused by a number of management staff of having 'founder's syndrome'. Those who were around us during those years would well remember the dramas of the 'attempted coup', 'organic-gate', 'pear-gate', 'An Honest Conversation', 'Food Connect 2.0' and, of course, the 2011 Brisbane floods. They would also remember winning the 2009 Premier's Award for Sustainability and the 2009 Banksia Awards, People's Choice Trophy – it was a smorgasbord of experiences.

FOOD CONNECT FOUNDATION

In 2009, after working with Nick Rose on a local food strategy for the Coffs Coast, the not-for-profit Food

Connect Foundation (FCF) was established to provide a structure for piloting replications in Australia, and to explore other areas of Food Sovereignty. Nick brought with him his recent experiences and awareness of La Via Campesina, and opened discussions for Australia's place in that movement. The foundation provided a space for me to explore other areas of interest such as farmland trusts, as a mechanism to protect prime farmland from continued urban sprawl; facilitating pathways for young farmers; and encouraging ethical 'foodpreneurs' to start small businesses of their own.

The foundation supported Think Food, a consultancy providing academic and practical advice in local food systems and rural futures strategies. It also provided one of the launching pads for initiating AFSA. With the Fair Food movement still in its embryonic stages, the period 2008 to 2010 saw many groups of people from all over Australia visit our homestead in Brisbane. They wanted to research our operations, and we gladly opened the doors and shared our information, processes and financials. Melbourne-based environmental education centre CERES initiated CERES Fair Food from this, and two groups from Sydney and Adelaide used our brand and software systems to set up separate enterprises. There was a lot of support to expand the model at this time, but the foundation wasn't resourced to support the replications. Additionally, the day-to-day pressures experienced by start-ups meant that member groups struggled to find the necessary time to work together. We've estimated that during that period, Food Connect Brisbane contributed around $200,000 to support the replication process. This does not include the work put into software development.

Interestingly, there was a lot of activity happening in parallel to our efforts. Approximately 15 other CSAs or local food-box systems independently started up and flourished after a visit to the homestead or getting hold of our business plan. We realised that my original idea of allowing the model to 'mutate', through open-sourcing all aspects of the business and taking a 'hands-off' approach, allowed the strengths of the model to be evolved. The lesson learnt here was that for a community organisation to be successful, it must develop its own social capital and resources unique to their region. Indeed, some of these operations were teaching us a thing or two, further emphasising mutual feedback loops that assist with adapting to new challenges and building resilience.

The foundation has been fully funded by Emma-Kate and me, as well as drawing on the limited resources of Food Connect, and so we've recently put the foundation's activities on hold due to a failed attempt to seek separate funding in 2013. We now have a volunteer advisory board who, recognising the imperative of keeping the enterprise healthy, provide advice and support to Emma, myself and the team. Besides supporting and involving ourselves in the activities of AFSA, the Open Food Network and the

Australian Food Hubs Network, Emma-Kate and I became founding members of the Slow Money Foundation based in the United States. The work of Slow Money is to facilitate a new type of investing, 'as if Food, Farms and Fertility mattered'. We consider the lack of financial investment in the new creative food economy a critical gap in the Fair Food movement, and hope to see an Australian chapter of Slow Money initiated soon.

KEY SUCCESSES

In 2010, a 'Social Return on Investment' evaluation performed on Food Connect estimated that for every dollar invested in Food Connect, $14.81 was created in social, environmental and economic impact. It was an impressive result, and we have since used this evaluation for communicating our impact. In broad terms, our model has created a number of areas of impact, as follows:

- improved farm gate returns, with an average of 40–55 cents in each retail dollar going back to the farmers and growers, as compared with 10–15 cents in the supermarket/central market systems
- improved health outcomes – Food Connect subscribers report an increase in their consumption of fruit and vegetables by as much as 50 per cent; and children of subscribers report a higher level of familiarity with fruit and vegetables

- being a source of inspiration for others around the country
- being regarded as a model of a sustainable social enterprise – reduced carbon and ecological footprints compared with mainstream food distribution.

The real success for us, though, are the relationships we have been able to foster between eaters and growers. Our farm tours and open-book policy for communication allow free and easy access between country and city. One morning, not long after I met Emma-Kate, she rang me to tell me how much she loved the baby-sized apples in the boxes. These apples would not have made it to the markets, but we stick with the policy of taking the whole crop. I immediately gave her farmer Joe Devereaux's number, and she called him to thank him. They chatted for over an hour. He rang the next day to say, 'That is why I get out of bed each day, Robert. I've got an extra spring in my step now, and you can be sure I'll put a lot more love into the next crop.' Emma still tears up when she recalls this experience.

It was also in 2010 that a wonderful pair of people approached me to discuss their organisation, Reciprocity. Dianne James and Dr Scott John had enjoyed long and successful careers themselves and were looking for ways to support La Via Campesina and its Food Sovereignty campaign, through building partnerships among farmers and farmer organisations, both overseas and in Australia. They donated two years of their time to conduct a listening tour of our farmers. Through their research, they've helped to strengthen our relationship with farmers and their 'reason for being' with Food Connect.

Since then, they've held regular farmer-centred events and Scott and Dianne have tirelessly provided support and encouragement to myself and Emma-Kate, and many of the staff at Food Connect. They've essentially become family. Scott has mentored me to trust my intuition, back myself more and communicate the original vision through developing a manifesto for Food Connect. Once the manifesto was written and communicated, Emma-Kate and I could use it to align our personal visions with Food Connect's overall vision. By this stage, we were encouraged by the momentum building in the United States with the local and regional food hubs, and incorporated these ideas into our big, hairy plans. Currently, Reciprocity is actively facilitating the Family Farmers United Network (FFUN), established by a cohort of past and existing Food Connect farmers and a number of other new farmers in the region. The network is made up of small-scale family farmers who aim to provide support for members to:

- maintain a stable and prosperous social, economic and cultural farming life

- share their ideas on natural farming methods aimed at developing improved and expanded natural and sustainable farming practices
- connect members with La Via Campesina to share with and learn from like-minded farmers around the world.

FFUN members are local (within a 300-kilometre radius of Brisbane), an individual or a group of people who grow food independently, without corporate monetary support. The group is currently meeting seasonally. A farmer member volunteers to host at their property and conducts a tour of their operations and share knowledge. Each visiting farmer is encouraged to bring seeds to swap, their favourite tools and new information from the previous meeting. They are building a level of solidarity completely different from many of the mainstream farmer networks in existence.

KEY LEARNINGS AND CHALLENGES

Bootstrapping the business has been Food Connect's main challenge. We often forget to remind ourselves that Food Connect was established with no start-up loans or grants from any source, which placed us at an economic disadvantage. Without significant capital investment, our survival has been due to tremendous personal and financial sacrifice and a deep commitment to the higher purpose, as well as the intangible but equally significant connection to our community of growers and eaters. We have sporadically been the lucky recipients of a few donations in troubled times, but rarely did they come with 'no strings attached'.

The living wage

We've often been criticised – and I would agree to a point – that while we may be achieving good social and environmental outcomes, our social justice outcomes for our team are more debatable. As stated earlier, we wrote into our constitution a 2:1 wage ratio, meaning the highest paid person cannot earn more than double the lowest paid person. Asking talented, committed people to accept living on award wages requires them to take a deep breath and re-evaluate that commitment, their relationship to money, and to enquire deeper into their value base and lifestyle choices.

With our wage ratio, we wanted to address some of the inequity between farm work and city salaries. In line with EF Schumacher's thinking in *Small Is Beautiful,* as responsible employers, one of our tasks is to seek ways to enable our people to flourish while protecting and conserving our resources and environment for current and future generations. At the time of our launch the average income a farmer earned was $8 per hour.

A living wage is defined as the wage that can meet the basic needs to maintain a safe, decent standard of living within the community. Using the National Awards, we could be both equitable and find ways to enrich our people's lives. We provide staff discounts on food, make our vehicles available, and find ways to support each other with sustainable lifestyle choices. Emma-Kate and I aspire to live simply, and we seek out working with people with aligned values.

While this has probably meant some lost potential talent when advertising positions, we are really proud of the team that has developed at Food Connect. We have seen low rates of absenteeism and turnover – despite being a highly part-time workforce – saved money in recruitment and training, established a wellness policy to address burnout, reduced reliance on welfare, increased social participation and fostered incredible loyalty and productivity.

Appealing to an invisible market

It's no mean feat to survive in a marketplace dominated by cultural assumptions of 'cheap food' and 'endless consumer choice' all year round. Changing the culture around food has been hard work – sometimes staff would sit on the phones for an hour explaining to customers why there was a heliosis grub in their corn. A substantial barrier to the expansion of the local food economy in this country is due to consumers conditioned to expect year-round availability and endless choice; for prices to remain 'down, down, down' for 'glamour' produce. One day I answered a call from a customer complaining about a grub in her broccoli, and I told her we should have charged her an extra $2 for added protein! It didn't go down well.

Food Connect did, for a short time, enjoy the benefits of being the first of our type, and we used that position to support the growth of the movement. Now, the market is starting to fill with mum and dad home-delivery food-box schemes, and many other alternative solutions that didn't exist ten years ago. This is a good thing! But it makes it even more challenging for us to remain viable because we experience the disadvantages of being the pioneer, like those coming after us having an easier time with set-up (as we've already blazed the trail), technological uncertainties, shifts in customer needs and the complacency that comes from not having competitors at first.

Spending vast amounts of money and time in creating software to meet our business rules has had an impact on our viability; however, the open source principle that we hold true meant we could share the code with other enterprises, enabling them to get started, and develop the software further. A good example of this is the establishment of the Open Food Network, which aims to provide open source solutions for the food revolution across the world.

The role of educating the market in order to create a new one has impeded our financial prosperity; however, the pure economic view ignores the initial intention: to create a new system, disrupt the existing one and make it obsolete. And, notwithstanding pending crises, we know the results will not happen in years, but generations. We also know that it won't come in the form of a 'Food Connect Empire'. The solutions must be unique to the environment in which the local community lives.

We need to ask the question: would people have left their corporate jobs and started their own food enterprises, farms, or food activism had organisations like Food Connect not been around ten years ago to nurture, facilitate and encourage them, or at least provide a demonstration that gave some hope of success?

One of our long-term subscribers and brilliant thinkers, Jason Grant, runs a design agency called Inkahoots. In 2012, because the major supermarket chains were increasingly undermining sustainable food alternatives (with their use and abuse of terms such as 'seasonal' and 'local', for example), Inkahoots proposed a campaign that would simultaneously resist co-option and function as a critique of corporate food systems. A series of 'contradictory' terms are now embraced by Food Connect: 'dirty', 'rough', 'imperfect' and 'unruly' as an antithetical badge of pride and a 'real food' ethic.

Choosing not to sell exclusively organic-certified produce has put us in conflict with the organic industry. The pay-off for us, however, has meant that many farmers respected us. We're not about excluding any farmers from our model. We're here for all farmers, and for all people who want to have fresh, nutritious food. We've seen a positive ripple effect, both within the farming community and the health sector, which often criticises organics as elitist and expensive. Our focus is to address the social justice issues for farmers – treating farmers fairly and making food affordable. It's been one of the most difficult challenges, but we are so proud of sticking with it.

I often get credited for championing the CSA movement in this country, but this romanticises things and doesn't acknowledge the many supporters I've had over the years. With this support, I've been given the confidence, the capacity and, more importantly, the *legitimacy* to effect broader social change. Food Connect as it exists today has been the culmination of many years of trial and error, a willingness to experiment, to fail, to observe and to learn from mistakes.

THE FUTURE

Recently, we've revisited our roots and expressed the desire to remain small but influential. As the retail sector is rapidly filling with alternative models and consumer demand is stepping up to the next level, we've realised that we need to look at our efforts in a different way.

Over the past three years we've diversified from just supplying fruit and vegetable boxes to individual households into wholesaling and value-adding. This was more in response to the demands from the market, but it also aligns with our vision to become Brisbane's Local Food Hub. This will give us increased exposure to different markets and increase the volume of produce for our farmers.

Humanity has the know-how to solve all of our problems. Contrary to the mainstream rhetoric, we have plenty of land, water and food. We just don't know how to share it. Our greatest challenge – in fact, the final frontier for humanity – is to get along with each other and share. If we fail at this, then humanity will fail. Through food being done fairly, I believe we have a great opportunity to provide a pathway for all of us to collaborate and realise our true potential.

Food has the potential to be the great unifier, the equaliser and mixed with some wine or 'spirit' around a table, can provide the catalyst for extraordinary events to occur. We debate and learn, we make big decisions, we fall in love, we discover new things about ourselves and the world. We eat at least three times a day, every day of our lives. Nothing else is consumed, discussed, felt, smelt and touched as intimately as food. When seen through this lens we can see the exciting potential in good food, done with love.

THE ACCIDENTAL FOOD SOVEREIGNTY ACTIVIST

MICHAEL CROFT

When I first started writing and talking about Food Sovereignty, in 2007, I knew that no Australian organisation was a member of La Via Campesina. It was always a personal dream of mine that this country would become directly linked to the global movement. So it's hugely satisfying for me on a personal level, but more importantly for this movement to grow in power and connectedness, that we now have a direct link to the global movement in the person of Michael Croft.

Michael was a pillar of support in the early days of building AFSA, serving as its first national president from 2012 to 2014, and playing a vitally important role in shaping the strategic direction of the organisation. I had the privilege of spending time with Michael and his wife, Elizabeth, at their family farm in Beulah, walking the contours and paddocks, admiring the herd of Belted Galloways and basking in the peace and quiet of that part of the country.

We've had long discussions late into the night over red wine and delicious farm-bred beef, about food politics and food movements, and the opportunities and constraints we operate within Australia.

I've watched with delight as Michael has entered the often confusing and challenging world of global food politics. It is hugely significant for AFSA to be participating in this space. If we're going to be effective in our aspirations for a transformed food system in this country, we have to fully understand and embrace the concept that we are part of a global movement of transformation. And that we need to work in solidarity with that movement, because its gains are our gains, and its challenges are equally ones we share.

I will always be grateful to Michael and Elizabeth for making the sacrifices that have allowed him to participate in those international arenas – which of course are not also without their compensations! And I'm very grateful to him for sharing this journey with a wider readership in these pages.

***

Whenever you find yourself on the side of the majority, it is time to pause and reflect.

Mark Twain

ROME, ITALY, OCTOBER 2014

I am sitting in a small room in Villa Benedetta, it's a beautiful autumn day, and I'm wondering how to explain how I got here, what being an international food activist means, and why I must do what I do. The expansive view from my fifth-floor window is of Rome's old central fruit and vegetable market, sitting as it does by the train line, and not far from the Tiber River. The markets are derelict but undergoing rejuvenation as an activity centre for youth. For me the buildings are symbolic of a food system that succumbed to the ravages of an ideology that places price above all other considerations, above culture, society, people and environment. That this should have happened in Italy of all places, the land where the sharing of good food and wine is truly a national pastime, is not lost on me.

In my youth and early adult years, being a farmer was not on my list of options. My family were not farmers, I had no relations who were farmers, and farming in my family was perhaps best seen as hard work we had left behind in our climb to middle-class respectability. At worst my peers perceived farming as hard and dirty work done by poor peasants who didn't have better options. So my journey to the heart and soul of international Food Sovereignty and peasant-based agriculture was as perhaps as accidental as it was logically sequential.

AINSLIE, CANBERRA, 1994

Ainslie was where my awakening began. I was 36 and, with my wife, Elizabeth, and two young daughters, had just purchased a fibro workers' cottage in the inner city on a 1500-square-metre block of land. Today this would be a million-dollar-plus purchase, but at the time it was purchased for well below the cost of an average house.

Our big urban block of land faced north and had fertile alluvial soils, both important and rare in Canberra. These facts nourished a desire to grow things and create our own oasis in the middle of the city. Not really knowing where to start, we consulted our good friends Jennie and Chris Curtis, who had a magnificent garden. Jennie is an award-winning landscape architect who uses Permaculture, and in no time we had a beautiful 'food forest' design to implement. This was complete with fruit and nut trees, vegetable mandalas, chooks, ducks and organic children who were now at Canberra's Steiner school. Much to our families' amusement we were evolving into those quirky, inner-city, bike-riding, chardonnay-sipping, salad-munching hippies.

Our children's education was important and the public school system was good if you were mainstream, but our daughters were definitely not mainstream. Consequently, the local Steiner school became a focal point in our children's lives and it was here that we discovered biodynamic agriculture. We were fortunate to be guided and inspired by the late Lynette West. Lynette developed her Certificate 3 in Biodynamic Agriculture using us as guinea pigs, the same course she later introduced to Prince Charles and his Duchy Farms.

Lynette's biodynamic farming course ran for many weeks and was an eye-opener. She gently challenged all I thought I knew about agriculture, food, plants, animals, biological functions and ecosystems. Here was a system of farming that paid attention, applied systemic analysis, relied on feedback, was closed loop, clean, intellectually challenging and, above all, made farming seem possible and rewarding to us naive city folk. Lynette's lumberjack cake was also a big hit and the best I have eaten.

As my family's education in biodynamics and farming was progressing, so too was our backyard poultry keeping. Two chickens became six, then eight, and then roosters and ducks appeared. We had been warned about the urban fox population by our growing collection of small-scale farming friends, and in an effort to prevent the premature demise of our menagerie, Lynette us told about Allsun Farm, their tools and Electranet fencing. After a phone call and some questions, Mike Plane of Allsun Farm arrived at our house and delivered 50 metres of electric net fencing and an energiser to power it. We think this was probably the first urban use of this technology in Australia – Mike was impressed and amused, our dog was not. Our dog's nose made contact with the fence shortly after it was erected, and she didn't leave the back deck for a week, so we knew our birds would be safe from foxes.

Our introduction to Joyce Wilkie and Mike Plane of Allsun Farm led us to their friend Joel Salatin, his books and his evangelical work of introducing farming to non-farmers. *You Can Farm, Salad Bar Beef* and *Pastured Poultry Profits,* Joel's only books at the time, joined our library. The seeds of an idea were germinating, but still below soil.

BEULAH, URIARRA, 2000

The new millennium was upon us, and our peaceful inner-city suburb was changing. The workers' cottages on the big blocks were being knocked down at an alarming rate. Flats, townhouses and dual occupancies were springing up everywhere. We felt surrounded and the poorly planned urban intensification was crushing our desire to live in a city.

Our youngest daughter, Briana, was by then eight years old and an experienced chicken expert. We had hatched them from eggs we had carefully collected and selected, we had grown them out, we had dispatched and eaten the excess roosters under the guidance of our octogenarian neighbours Wal and Zeta. At some point Briana announced, 'I want to live on a farm.'

This suggestion was not immediately dismissed. A decade or so earlier, Elizabeth and I had flirted with the idea of buying an acreage block of land on the outskirts of town. On a couple of occasions we almost took the plunge, and then the reality of the long daily commute to work that would make us environmental vandals would surface. Enter Beulah, the promised land of *Pilgrim's Progress.*

Beulah is a 200-acre small farm in the foothills of the Brindabella Mountains west of Canberra. I stumbled upon it while searching a new real-estate website for friends who were moving to Canberra. Our internet search did not result in love at first sight. The farmhouse looked like a cheap motel in the photos, the description left a lot to be desired and many questions remained. On the plus side, the asking price for the farm was about what we thought our urban home was worth. It was also only 20 minutes by car from the Steiner school our children attended, and a similar distance to work. There was a school bus close by, so our carbon footprint would not increase. We decided to investigate.

The drive to the farm was through an extensive mature pine forest and farm land. There were very few cars on the road. As we progressed west from Canberra, it got greener. As we walked the property, we began to realise there was infrastructure, dams and sheds. By the time we had toured the property and seen wallabies, wedge-tailed eagles, wombats and an echidna, we were sold. The motel-like house, well, we could deal with that later.

We were very excited; however, we needed to rationalise our emotional purchase of the farm. We told ourselves we were just 'moving house', we'd have more space for our children and their friends, we could transplant the fruit trees from our fruit forest, the chooks and dog would love it, and the property developers were literally closing in on us. Besides, we weren't going to be farmers; we would have other farmers put their stock on the land, keep the grass and fire risk down, maintain the fences and pay us a small fee in cash or kind. This was going to be a simple and uncomplicated move, and it was – until we got there.

I don't think you ever plan on falling in love, it just happens, and it happened to me. The farm gave me an unexpected, welcome and rare gift; it put both literal and metaphorical distance between the noise of the city and me. The farm gave me the most precious gift of all in a hyper-connected world: the space and time to think and listen.

For what seemed like an eternity I dreamt of quitting my work and farming for a living – how hard could it be? I already had the farm, I'd read the books, I was good with animals, we were in a high rainfall area, we had great infrastructure and enough water storage to confidently say the property was droughtproof. I was doing every farming and farm-related course possible; my farming library was becoming quite extensive. Okay, I admit it, I was obsessed and in love with the generosity of the farm and the environment that sustained it. Elizabeth knew I was obsessed and eventually gave us her blessing to have an affair.

Reality always bites back, of course. I remember the exact words of my fourth- and fifth-generation family farmer neighbours. To a man, and they were all men, they said, 'Mate, you can't make a living off 200 acres. Not in this country anyway.' I didn't react particularly well to being told I couldn't make a go of farming on my land. A certain amount of male pride and arrogance took hold: 'I'll prove them wrong.' It was only days later that I started to wonder what made these experienced farmers say and believe that – and what if they were right?

My curiosity was aroused, and, if my farming neighbours were right, I wanted to know what was wrong with the food and agricultural system that would make this the case. Little did I know 14 years ago that this seemingly simple question would lead me all the way to the international struggles against a deeply dysfunctional industrial food system at the United Nations Rome-based agency the Committee on World Food Security. At the time I thought, 'If the system doesn't allow me to earn a living from 200 acres, the system is flawed. So I have two choices: either I change the system or I find a way around it.' Since I didn't know how to change an entire system, I took the easy option and circumvented the shortcomings and injustices of the existing system.

CIRCUMVENTING THE SYSTEM

I reasoned that if farming conventionally in my region was not profitable, I needed to do everything differently. My farm's Unique Selling Proposition (USP), to use a business phrase, was going to be many USPs combined. So everything we did was deliberately unconventional.

The farms around us are fine Merino wool farms that also run a few cattle on the side for income diversification. The logic being if wool is not doing so well in the market, you can always sell a few cattle because people always have to eat. This was the linear logic of an input-output, almost mono-cultural farming system with a small safety net of sorts.

I have always had a way with animals, so we decided to be a multi-species farm where each species occupied a different and complementary ecological niche, preferably using the same land while enhancing the fertility – Permaculture and Joel Salatin's influence made manifest. It was also spreading the risk in case one or more species proved problematic. Ultimately, this would mean multiple income streams from the same land, known as species stacking, and if we farmed using biodynamics, we would have significantly less inputs and the products would command a premium. So we aimed for more income streams and higher prices. And it worked. We ran sheep, cattle, pigs and chickens on the same land sequentially. We had beehives working the same land, and more species were added later, working dogs and llamas for stock protection, and three species of fish in the dams.

Very early on in my research on how to make farming pay, I came across the work of Allan Savory, which was then called Holistic Resource Management, now Holistic Management (HM). HM made intuitive sense to me, and I wanted to apply it to increase the carrying capacity of the farm and sequester carbon in the soil simultaneously. It sounded like a win-win, but I feared 'a little knowledge was a dangerous thing'.

Synchronicity has a strange way of appearing when you need it. I was holidaying at the snow and wasn't skiing due to a knee injury, so I was left reading Allan Savory's book *Holistic Management.* At lunch I was introduced to a grazier; in quick time I asked him, 'Have you heard of Allan Savory's work?' and he replied that he had. I grabbed the book and showed him, enthusing that I suspected it contained the answer livestock farmers and climate-change researchers had been looking for: livestock sequestering carbon, improving the soil, increasing carrying capacity and saving the planet. Brian Marshall smiled kindly and told me he knew this; he turned out to be an HM educator and was one of those responsible for first bringing Allan Savory and HM to

Australia. Brian has since become a friend and soon after our meeting ran an HM course on our farm for interested farmers. HM works, we have used it extensively on the farm, and it enabled us to 'grow' topsoil during the six-year Millennium Drought.

So now we were growing animals for food, really good food, and it had come time to sell our products. I found that the retail price of food was acceptable, but only for those who sold retail. If you sold to the wholesalers, you often received less than 10 per cent of the retail price. As we had read Joel Salatin's work almost a decade earlier, his philosophy of innovative direct marketing and cutting out the wholesalers made perfect sense. So we started selling direct to family and friends, and this worked. At about the same time the farmers' markets were developing in Canberra, so we started going there every weekend and selling at retail prices there too. Life was good – busy but good.

Our animals were chemical-free and organic, but I decided that they needed to be different and then some. In my research of suitable breeds I had come across the alarming facts that we were losing some three to four breeds a month globally to the homogenising demands of industrial agriculture. This dramatic loss of biodiversity is the same if not worse for edible plants where in the last 100 years we have lost 75 per cent of food plant diversity. These shocking figures set alarm bells ringing for me, as my then basic understanding of ecosystems was that they were only resilient as both the breadth and depth of diversity in them. Consequently, we started farming rare breeds as part of an in situ conservation program.

I became increasingly vocal about this loss of our food bio-diversity and associated issues, talking to all my customers, the radio and press. In sticking my head up, I got noticed, and I was co-opted onto the board of the Rare Breeds Trust of Australia. The trust does excellent work preserving animal genetic material, and it is not dissimilar to the many heirloom seed-savers networks around Australia and the world.

Shortly after this, our farm, called Mountain Creek Farm because it is on Mountain Creek Road, came to the attention of the local Slow Food chapter. They invited me to their first 'Taste of Slow' in Canberra as a vendor in 2006; our success at the markets meant that we had no product to sell, but we went anyway to spread the word about a dysfunctional food system that squandered resilience by reducing food diversity. It was early days for Slow Food's Ark of Taste in Australia, and I was co-opted onto the Ark of Taste as well.

Slow Food is often subject to criticisms, most tall poppies are, but I am genuinely grateful for their support and for introducing me to many concepts and interconnections, including Food Sovereignty. My association with Slow Food exposed me to a world where food, the systems and the people that produce it really matter. Slow Food recognised the significance of my farm's work; as a result I was the region's first delegate to Terra Madre in Turin, Italy, in 2008. Terra Madre is an amazing experience of hundreds of small-scale food producers gathered to share their stories and raise awareness – well worth going if you get the chance.

Our efforts to be different, and prove that you could make a living on 200 acres in temperate Australia, became increasingly recognised. Our produce featured in many high-profile events, our pork was chosen as part of the relaunch of the Sydney International Food Festival. We were constantly turning away chefs who wanted to feature our produce, and enquiries came from all over the globe. We were the region's meat producer of the year in 2009. In short we were successful and had become 'a legend in our own lunch box'.

But ... I wasn't happy. I loved farming and food, but the system was ill and the symptoms looked terminal. The more I beat the system at its own game, the more intimately I understood the system's flaws. These flaws trapped my fellow farmers in a cycle of debt and near poverty from which others benefited handsomely. I came to understand that the system was parasitic and unjust. It simply wasn't fair, and giving people a fair go is quintessentially Australian. My social justice button was being repeatedly pushed, and there was no off switch.

CHALLENGING THE SYSTEM

I was doing a lot of public speaking by invitation; community groups are always looking for someone to speak to their members at gatherings. I was linking the systemic dysfunctions of agriculture to societal and environmental ills in interesting ways, and I started to gain a broader audience. Invitations came from the local public radio broadcasters after I started using variations on Wendell Berry's famous saying 'Eating is an agricultural act!' I used 'Eating is an environmental act!' and later 'Eating is a political act!' as headings for talks, which captured both the connections and complexities of the issues.

Early on I had started identifying and linking the individual symptoms of food and agricultural dysfunction in my talks. Individually, these symptoms seem like issues that can be resolved with a policy tweak here or there. However, when linked together they form a damning picture of a food system in crisis, and one the Australian government and industrial agriculture and food would prefer not to mention.

The list of inconvenient truths is sickening, and shows the deep dysfunctions that are the hallmarks of an unsustainable food system. I defy anyone to read through the list and tell us all that Australian food and agriculture are not in crisis. And yet this is what many at the highest levels of government, academia, bureaucracy and farm lobby groups do repeatedly. Temporary market conditions are usually cited as the problem, but six decades of data show they aren't temporary – they are systemic and permanent until collapse occurs.

The inconvenient truths of Australia's food system

- Farmers suicide at twice the national average rate.
- Seventy per cent of farms are only viable due to off-farm income.
- The average age of farmers is close to 60 and slowly climbing.
- Fifty per cent of farmers hope to retire in the next five years.
- There are four jobs for every agriculture graduate, but the agriculture colleges remain stubbornly empty.
- In 2013, for the first time ever, banks owned more than 50 per cent of all farm equity in Australia.
- We are losing 76 farmers a week, on average, from a low base of some 140,000 or 0.6 per cent of the population.
- Biodiversity loss is at record rates and, since agriculture covers 53 per cent of Australia's landmass, agriculture is largely responsible.

• The terms of trade for Australian agriculture have been declining for 60 years and continue to decline, according to the Australian Bureau of Statistics.

• We have only one tomato cannery left in Australia, and healthy vegetable and fruit crops are being routinely ploughed in as 'no longer viable'.

• It takes ten calories of oil to produce and deliver one calorie of food, which is criminally inefficient if not negligent.

• The three daily meals Australians eat have travelled well over 10,000 kilometres before they reach a stomach, with the 25 most common weekly food items in a Melbourne shopping trolley having travelled 70,803 kilometres.

• Almost 80 per cent of the retail grocery market is controlled by two corporations.

• There are only two days' worth of staples and five days' worth of all food on the supermarket shelves at any one point in time, yet we are considered food secure.

• As Australia is 'open for business', foreign transnational corporations increasingly own our food production, distribution and export networks.

• For the first time in history, children born in Australia today may have a shorter life expectancy than their parents, as their eating habits are fuelling an obesity and diabetes epidemic.

- It is a national disgrace that, as an agricultural exporter that can feed three times its population, in 2012 Australia had 2.265 million food-insecure people, and 575,000 of these were children under the age of 15. Food-insecure means they don't know where their next meal is coming from and go to bed hungry one night a week.
- We waste nearly 40 per cent of all the food we produce, mostly thrown away as post-consumer waste.

I am an optimist by nature, and it really it pains me to be the bearer of bad news – but there's more.

To prove the point about niches, farmers' markets (themselves a niche) are restricting the number of vendors who can sell certain niche and specialty products, because there would be too many sellers for those niche products to remain viable.

Then there is the laudable goal of every Australian government and council of ending food waste and loss. It breaks my heart to think of my friend who ripped out his fruit trees and attempted suicide (you can watch a short documentary on his story at https://vimeo.com/71938049), or even someone who allows my produce to spoil and throws it in the rubbish bin, but what happens to the prices farmers receive if we manage to eliminate food waste and loss without addressing the system's dysfunction as a whole? Some

30–40 per cent more food would be available for consumption at no extra cost, as it has already been paid for. It is a wicked problem when fixing it creates more problems. Wicked problems are merely elements of a systemic problem that are being addressed in isolation.

So all this begs the question, in whose interest is it for farmers to increase production in a global food glut? Who is calling for a 70 per cent or greater increase in food production, while globally we waste 32 per cent of all food, and why are they saying this will feed the world's poor and save farmers? A farmer in the Global North growing more maize that is unaffordable to someone from the Global South won't help. People of the Global South waste very little if any food. Yet farmers are being called to do the same thing they have always done, increase productivity – which in the aggregate means increase production – and expect a different result. This is one clinical definition of insanity, so are we insane?

Food gluts and niche farming

- According to the United Nations, the world produces enough food already to feed 12 billion people, so why are there calls to increase production when the global population is 7.2 billion?

- There is a global glut of food, which also helps to explain why farmers are price takers (rather than price setters), and power imbalances and market

> distortions merely compound this problem for farmers of a surplus-producing exporting nation like Australia.
>
> • Smart Australian farmers target niche markets, and niches are a market response to a glut.
>
> • As soon as a niche is identified, other farmers rush into it, lowering prices and turning the price-setting niche market into a price-taking one.

For nearly a decade I felt like a voice in the wilderness. Of course I realised that I was not alone, I had sympathisers and comrades, but they were the already converted few. I initially (and naively) felt that if people knew the facts, things would change because we have 'common sense', and the food system didn't make sense. Increasingly, I came to realise that common sense wasn't common, and I hadn't bargained on the wilfully blind, and the power of the paradigms and memes that support them.

AUSTRALIAN RURAL LEADERSHIP PROGRAM AND FOUNDATION (ARLP/ARLF), 2011

I was fortunate to be a participant on course 18 of the Australian Rural Leadership Program, and it was transformational. It is ironic that you need to go on a deeply personal and individual journey to discover that we are all connected and in this together, and that it is only as a collective that we solve systemic problems. So it was for me.

In Australia we still talk of community, but there have been decades of atomisation and individualism encouraged by a system that does not want us to organise collectively for the common good. The last remnants of community and collective action are predominantly found in rural and remote Australia. The various rural fire services, crewed mainly by volunteers, are obvious and vital manifestations of community spirit, but in the main a sense of community has been deliberately weakened. It is not in the interests of the system, and the few that profit most from it, for people to decide collectively what is in their best interests. Successive government policies have aided and abetted the divide-and-conquer drive of individualist consumerism. In fact, collective action is really only tolerated in times of crisis such as bushfires or floods, when individual private interests are obviously inadequate and can't cope.

AUSTRALIAN FOOD SOVEREIGNTY ALLIANCE (AFSA), 2011

My friend Nick Rose has written about the formation of AFSA, but Food Sovereignty was an idea whose time was well overdue in Australia. Food Sovereignty was my 'aha' moment, the moment when all the pieces fell into place and the food system's many parts came into focus. It seems strange that only four years ago, the term Food Sovereignty evoked confusion among food activists in Australia, and only a few academics and researchers knew what it meant. AFSA changed all that.

Many food activists still speak of the need for 'food security', but I would caution against using that term, and here's why. An African-American urban farmer, feeding his community in Oakland, California, was asked by Eric Holt-Giménez, CEO of Food First in the United States, 'So tell me, what do you think the difference between food security and food sovereignty is?' The farmer immediately replied, 'Dude, I can be food secure in jail.' My own abridged distinction is this: food security means you are dependent on someone else's food system to keep you fed, and food sovereignty means it's your food system. Or my ultra-short version: food security is dependence, food sovereignty is freedom.

I was deeply honoured to end up as president of the fledgling AFSA, and we had a huge task ahead of us. That we are here today collectively writing a book on Food Sovereignty in Australia speaks volumes. I would, however, strongly encourage you to visit the AFSA website and see what we've achieved as a 100 per cent volunteer organisation.

LA VIA CAMPESINA'S SIXTH INTERNATIONAL CONFERENCE, INDONESIA, JUNE 2013

It was an honour and privilege to be asked to attend the international conference of the world's most influential farmer-led, farmer-only member organisation. It was at this four-day meeting that I found my tribe.

As a farmer afforded all the privileges of one of the richest nations of the Global North, I found commonality and solidarity with peasant farmers from the poorest nations of the Global South. We are different, of course, and yet our struggles are the same. Not just the physical farming issues of weather, climate, pests, diseases, and so on, but also the political struggles against the neoliberal hegemony that would see all small-scale family farmers eliminated from the system.

That long list of systemic dysfunctions in Australia's food system is found the world over. As globalisation increases, the problems farmers face all have a common denominator – the sociopathic logic of capitalism. My fight for a fair, just and sustainable food system is little different from that of the peasant farmer's fight in India, Africa, Latin America or Asia, or the small-scale family farmer's fight in Europe, the

United States, the United Kingdom or Canada. We are all in this together.

In Jakarta I made wonderful connections, many friends, and experienced the great outcomes that a global movement based on solidarity can and does achieve. The logistics for this meeting were impressive with some 600 delegates attending from around the globe and simultaneous translations into four languages. The 100 or so translators, interpreters and technical assistants were all volunteers, and they spent the days and nights facilitating discourse between the different linguistic groups. Just imagine getting 183 organisations from 88 countries, complete with language and cultural differences, to agree to unified position statements and policies. Achieving genuine consensus in these circumstances was very impressive, the goodwill was palpable, and it was a wonderfully visceral introduction for me – and a formal introduction for AFSA – to the international Food Sovereignty movement.

It was in Jakarta that I was buttonholed by a few senior members of La Via Campesina. They heard me speaking about the systemic dysfunctions of the most industrialised agricultural nation on earth. They nodded in sympathetic agreement and also indicated that I wasn't telling them anything new – solidarity. I was asked a large number of probing questions, and I was being assessed. Suddenly, I was interrupted by a gregarious, obviously experienced and knowledgeable Italian, who said emphatically, 'We need you in Rome!'

What did he mean? What use would a small-scale family farmer from Australia be there? Andreas repeated, 'We need you in Rome!'

UN COMMITTEE ON WORLD FOOD SECURITY, CFS 40, ROME, OCTOBER 2013

If you don't like acronyms, then being on the CC of the CSM of the CFS of the UN at the FAO will be no fun at all. Let me expand that statement. I am currently the Australasian focal point delegate member on the Coordinating Committee (CC) of the Civil Society Mechanism (CSM) of the Committee on World Food Security (CFS) of the United Nations (UN) at the Food and Agriculture Organisation (FAO) in Rome, Italy.

The La Via Campesina people had been rather insistent that they wanted a Food Sovereignty person from Australia in Rome, and so too had others I met from the International Planning Committee for Food Sovereignty (IPC). It was disconcerting to be told by experienced social movement leaders that Australia is considered part of 'the evil triumvirate' as far as small-scale food producers are concerned.

Australia, Canada and the United States are the dogmatic pushers of their drug of dependence, global free trade. Their self-interested dogma threatens hundreds of millions of families with poverty and destitution. Australia in particular is seen as a selfish, uncaring ideologue by global social movements, and this was before our lurch to the right under the current federal government. I was briefed that an Australian presence and voice in Rome was needed to help counteract Australia's arrogant assumption that we knew what was best for peasants of the Global South. Upon my return from Jakarta I investigated what I could do to help in this global justice issue.

It turned out that the existing Australasian focal point delegate had resigned due to ill health and that the position was temporarily vacant. I immediately contacted those in Australia that had been involved to see what I needed to do – as the CSM renewal process had begun and election was required. As I had been briefed in Jakarta, and then researched the issues for three months, I was permitted to nominate and was subsequently elected. Australasian in this context means Australia and New Zealand.

My role at the CSM is to input the messages civil society has about food and agriculture in Australia and New Zealand to the CSM and to be part of the working groups on areas of interest, speciality and concern. I then inform civil society in Australia and New Zealand of the proceedings and outcomes of the CSM and CFS.

Initially, I felt that I was failing spectacularly in fulfilling my new role, until experienced CSM people told me that it takes at least two to three years to understand the process and become effective. The CSM is a totally new mechanism within the inter-governmental body that is the UN, and the only one within the entire UN system that permits direct input and participation from civil society.

The UN is a labyrinth of rules, procedures and protocols that takes years to master, and the CSM is still working out internal protocols that will satisfy over 900 non-governmental and civil society organisations from around the world, not to mention multi-donor trust funds and the UN itself. Throw in simultaneous translations into eight languages and cultural differences, and it is a complex and demanding volunteer position that receives little thanks and a whole lot of criticism.

The CFS is the political body for food security within the UN system, which is why it reports to the UN General Assembly. Being the political body means the CFS is where policy guidelines are developed that countries agree to and adopt. If you think Australian politics is contentious, try getting over 150 countries to accept and ratify policy guidelines that they have to take home and justify. Nine days straight of 15-hour meetings and high-level negotiations are not how most sane people would choose to spend their time in Rome – double-shot espressos help!

IPC MEETING, BRAZIL, NOVEMBER 2013

I was asked at the CFS meeting in Rome if I would be available to represent Australia and AFSA at the IPC meeting in Brazil – they liked what AFSA was doing and what I was saying. I said yes but, as a social movement, AFSA had no money for this type of activity. I had run out of personal funds at this stage. Fortunately, a sympathetic government had chipped in money for Global South countries, which in turn freed up some funds for an Aussie, so the trip to Brazil was funded.

The IPC meeting was fascinating; about 40 of us gathered to discuss the strategic directions for 2014 and 2015 for the global Food Sovereignty movement. We came from every continent and translations into the 'colonial languages' of French, English, Spanish and Portuguese were required.

Obviously, land grabbing was high on the agenda, but so were the TTIP (Transatlantic Trade and Investment Partnership) and TPP (Trans Pacific Partnership) mega deals, and FPIC – Free and Prior Informed Consent. These discussions are then reflected in policy positions the global Food Sovereignty movement takes to the CFS meetings and works on in each region.

This reinforced for me the importance of AFSA's involvement in the international struggle for Food

Sovereignty for several reasons. First is because we are not alone and there is strength in numbers. Second is because we are collectively influencing the debate at the highest level. Third is because by sharing our experiences we learn what works or might work in our own regions. We also learn how to communicate and collaborate effectively across cultures.

I'd like to share just one of these communication 'learning experiences'. It was late on the second of our 14-hour days at the IPC. I was paying attention to everything, nodding in agreement, taking notes. I was in full agreement with everything being said, and my body language showed it. As I could add nothing to the discussion I was being respectfully quiet, as is customary in Australia. In truth I felt overwhelmed by the experiences of farmers in Latin America. One woman who spoke had lost two sons, a brother and a husband, killed in their struggles against land grabbing. What could a Global North farmer possibly add after hearing that?

At our last coffee break for the day, an experienced IPC member came up to me and said somewhat abruptly, 'What is it you don't like or disagree with? You have said nothing.' Enter cognitive dissonance. 'No, no,' I protested, 'I agree with everything that was said and can add nothing.' She replied, 'Well, you must say something! To say nothing is worse than disagreeing. To say nothing is to withdraw from the process and invalidate it.'

So there I was thinking that I was being agreeable and supportive, while many around me thought I was invalidating the entire process. Fortunately, I had several opportunities to speak and redeem myself. I must have done okay, because I was asked to give the formal thank-you speech at the close of the four-day meeting.

The next IPC meeting is in Mali, Africa (Ebola permitting), in 2015 and on agroecology. AFSA will be invited to attend the event and, funding permitting, I will be there to contribute in solidarity.

FAIR FOOD FARMERS UNITED (FFFU), MARCH 2014

The desperate need for a national farmer-only organisation had been much discussed in food and agriculture circles in Australia. The discussions on food and agriculture had been piecemeal and treated individual symptoms as if they were causes. Early in 2014 the AFSA committee met at Jonai Farms in Victoria for a strategy-planning weekend, and, after lengthy discussions over dinner and a fine bottle or two of red wine, Fair Food Farmers United was born.

Several needs were identified, one being the need for peer-to-peer sharing among farmers that would enable agroecological solutions to be promoted based on actual farm experiences. Another was the need for farmers to represent themselves to all levels of government and not via third parties. Both are needed and both are massive tasks to undertake at a national level.

It is early days for the FFFU, and much will happen as the critical membership mass is reached. This is happening fast, and if you know of farmers who are questioning the system that they are stuck with, and who are tired of bandaid solutions, please suggest they contact the FFFU – many hands will make light work.

CFS 41, ROME, OCTOBER 2014

Again I attended the CC, CSM and CFS meetings, this time as a 'fully funded participant'. I was still a volunteer, but at least airfares and basic living expenses were covered. Funding came from a multi-donor trust fund, which includes countries, sovereign wealth funds and big international NGOs. This time around I had a much better understanding of what was going on and why, I was also contributing more, but I was still learning – what were 'friends of the chair' meetings?

In the lead-up to CFS 41, working groups formed around the topics for decision boxes, and I joined several. One was 'food loss and waste in the context of a sustainable food system', another 'responsible agricultural investment'. These are highly political and contentious issues to get some 150 nation states plus the CSM to agree to – 'agreed language' is important.

You haven't lived until you have batted one small sentence back and forth to various nation states for an hour, only to finally agree on the original wording! Now imagine doing this for a few hundred sentences so they can end up in a UN policy guideline decision box, and you get the idea of a 'friends of the chair' meeting. This negotiation style usually results in 18-hour days in windowless rooms, locked in negotiations with the representatives from many countries, all protecting divergent national interests, while trying to agree to guidelines for the common good to end world hunger. Sounds like fun to a farmer, doesn't it?

Seriously, though, the CSM has a reputation for being tough negotiators in these meetings – and we are because we are fighting for the common good. We have had many wins and a few losses. An example of a win is that the CSM put agroecology on the agenda of the UN. Agroecology is the child of Food Sovereignty, and France has come out strongly in favour of agroecology as a means of combating climate change. Food sovereignty is also entering the lexicon of the UN courtesy of the CSM.

The important thing about the CSM is that it scares nation states – yes, the nation states bluster a bit, but they fear civil society, and rightly so. The sole mandate of the Committee on World Food Security is to end world hunger, and pursuing national interests usually clashes with this mandate. Civil society does not hesitate to tell nation states when they are wrong and being counterproductive.

Involvement in food activism at the UN, and the advancement of the agenda are important, but – and it is a big but – they are nothing without the continued grassroots activities of the local food movements at home. I see the grassroots activists at the local level and food system activists at the intergovernmental level as a strategic pincer movement. If you will forgive the military analogy, we are equal and supporting flanks in battles for the common good. Our enemy in these battles is the status quo of industrial food systems, and the organisations and governments that support them.

A lot has been written about the co-option of social movements by the state and authorities, with the premise being you can't be an insider to the system you want to change. As the Haitian saying goes, 'Unity of the roach and chicken happens in the belly of the chicken.' So what is a social movement to do if the state co-opts or suppresses it before it can threaten the status quo? I find myself engaged with the system I want to change at the highest level, while simultaneously being part of a grassroots social movement, but there is no road map. There is no precedent for maintaining grassroots integrity, values, principles and solidarity when engaging formally with the UN. We must stay true to our roots as we engage the system, drawing the map and making the road as we walk it.

I know from firsthand experience that La Via Campesina, the IPC and the CSM have fiercely 'held the line' in all the highly political UN policy guideline negotiations. More than once we have publicly condemned the policy guideline conclusions we were engaged in developing – the end result wasn't good enough. We then offered counter proposals and interpretations of the guidelines that work for social movements. I also know that the policy working groups, where we develop our positions prior to negotiations, are open, consensus-driven, collaborative spaces.

We have had losses, of course, but we have had significant wins that would not have happened were we not 'at the table'. All in all, I consider the upside of being engaged with the UN outweighs the value of remaining the vocal outsider, but can we be both? I am working to ensure that we can be both, but I am also acutely conscious that we should be forever vigilant that this is not the beginning of the digestive process in the belly of the beast.

The highly vocal outsiders of the grassroots are comrades to those who are being vocal on the inside – we act in solidarity as we perform different tasks. I sometimes use the metaphor of a flowering plant to describe what we are doing at the UN. Social movements are agitating and changing the system at the roots, while social movement delegates are the stems and branches introducing the energy of the roots to the entire plant to enable it to flower and set fruit.

A PARTING THOUGHT (OR THREE)

There is not much freedom to 'speak your mind' at the international negotiating level, mainly because you speak on behalf of others – in the case of global hunger and food rights, you speak for billions of people in real need. These committees and meetings are the living embodiment of the saying, 'There is no "I" in team.' There is little glory either, as the struggle is and will remain a long one, but there is a deep satisfaction in working on social justice issues for the common good, and putting others first.

It also takes a lot of time to know who the main players are in all these organisations, the informal and formal rules of engagement, protocols and procedures, committees and sub-committees. Understanding why things are the way they are takes longer still. Within social movements, personal contact and trust still counts for much. However, I can confidently say that through AFSA, Australia has been fully accepted and warmly welcomed as part of the international movement. This is a process that usually takes many years, but in our case, due to intensive engagement and the desire of the IPC and La Via Campesina to recover 'the lost continent' of Australia, this has been achieved very quickly.

It also costs money to engage in international food activism, as it is all volunteered work and unpaid. Travel is sometimes funded, as are accommodation and meals, but not always. NGOs do help fund people from Global South social movements to attend meetings, but not those from the Global North. It is assumed that if you are from the Global North, you are rich and can find funding. This might be true if you weren't trying to change the status quo, but social movements the world over are always cash poor. If you run a farm and have no one to do the work when you are away, and you do voluntary work for a social movement without funding, it is really tough going.

The international linkages and the solidarity they provide are of great importance to those in the Global South. Perhaps they do not appear to be of much use to Australia or New Zealand at first glance. However, the reason it is important for Australia to be represented at the CSM and the IPC is that Australia in particular calls for increased neoliberalisation of the global economy – more free-trade agreements and to let the markets rip. Being a counter voice to our government's ideological obsession, self-interest and greed really does help the entire movement. I am immensely honoured to do this work on behalf of both Australian and global civil society.

I'll finish by saying I don't like leaving the farm, family and friends, and travelling to Rome to do jet-lagged battle with the institutionalised stupidity of the world's governments. It is certainly no picnic or junket, and being a social movement activist is the same the world over; your commitment to the cause comes at personal and financial cost. It is hard work that must be done, and fortunately this is offset by the friendships and global solidarity you experience as you work for the common good. It is deeply satisfying when your social justice values and your work align; this is why I do it. I may well be an accidental activist, but my values leave me no choice. In solidarity!

THE TOXICOLOGIST TURNED FOOD FOREST MAESTRO

ANGELO ELIADES

I met Angelo in 2012 when we were searching for outstanding exponents of urban agriculture in Melbourne. Angelo's name soon came up, and not without reason: what Angelo has achieved in his small suburban backyard in Preston is nothing short of extraordinary.

Angelo has conclusively established what he set out to do: prove that a Permaculture food forest can be a highly productive, water conserving, soil-and-biodiversity enhancing, and largely self-sustaining form of urban food production. He has shown what many people have doubted – and many continue to doubt: that food production in cities, when undertaken efficiently and by knowledgeable and capable people and organisations, can make a significant contribution to local community food needs.

Just how significant that contribution might be remains to be seen, but if the experience of Havana during Cuba's 'special period' in the wake of the collapse of the Soviet Union in the early 1990s is anything to go by, it could be very significant indeed. And in a time of increasing uncertainty, with our globalised and industrialised food system exposed to climate shocks and the real likelihood of ever more expensive chemical and fossil fuel inputs, a skilled and capable urban-agriculture labour force could be an extremely important asset for any city in the world to possess.

What struck me most when I met Angelo was his generosity in sharing what he has learnt with others. That spirit of sharing and generosity is very evident in this extraordinary chapter, in which Angelo shares vivid and at times painful memories from his childhood, and weaves this into the philosophy and practice of the food forest that has now become his *raison d'être.*

***

> *I teach self-reliance, the world's most subversive practice. I teach people how to grow their own food, which is shockingly subversive. So, yes, it's seditious. But it's peaceful sedition.*

Bill Mollison, Permaculture co-founder

A RELUCTANT BOY GENIUS

Sometimes small and apparently insignificant events can trigger huge changes in our lives, much like the way the metaphoric flapping of a butterfly's wings can trigger a hurricane in a distant place several weeks later. Unlike destructive hurricanes, though, the outcomes of such events can be very constructive, charting us on a course in life that we never planned, pushing us into experiences that lead us to profound personal growth, prompting us into action that brings about positive change in the world around us. Mine is such a journey, and it started a very long time ago.

I was a very curious child, preoccupied with understanding how things worked. I could almost say I was born with a predisposition to scientific thinking. As you can probably guess, that meant I was a fairly introverted person as a child. I actually avoided public attention, but despite my best intentions, it came to me at a very young age.

No four-year-old plans on being on television, and the experience is extremely overwhelming. I still have fragmented memories of the cameras, bright lights, strange people and how distressed I felt. I recall meeting the TV celebrity and host Philip Brady, whom I used to watch on television. He took me to the staff canteen in the studio and bought me an ice-cream to make me feel better. I just wanted to go home. This was the first time in my life I had to perform for an audience, and, frankly, I hated it.

It didn't end there as I'd hoped. Photographers and journalists from more than one newspaper visited me. As I look at one of the yellowing newspaper article from over four decades ago, which my mum dutifully kept for me, I am reminded of what all the attention and fuss was about.

On a Sunday in early March 1973, the *Sunday Observer* newspaper ran a human interest story, titled, 'Brian Blackwell meets a junior genius, Angelo knows all the answers.' The subtitle boasted, 'Angelo Eliades is only four years old – but already knows more than most grown-ups.' The journalist discussed how he quizzed me about various world leaders, then switched to history – who invented the steam engine; who discovered electricity? – and how I fired back the correct answers without any hesitation. He went on to say that he saved the best till last. 'Geography is Angelo's best subject. He can reel off all the world's capitals from Reykjavik to Algiers. The whizz kid's favourite companion is a book of maps of the world.' The article explained I knew the alphabet in both English and Greek, and was learning to read and write in both of those languages.

The truth of the matter was that I really loathed geography; it was my father's favourite subject, not mine! To me it was rote memorisation and nothing more. My real passion was learning about all the wonderful things that surrounded me on this amazing green planet. Later in life I would discover that other people on this planet shared this passion to understand the natural world and called it *science.*

It was inevitable that I'd have something to do with the natural world in my future, but if anyone had told me that I'd pursue a career as a public speaker, a presenter and a trainer, standing in front of audiences, I would not have believed them. Life does take some strange turns.

People often ask me if I grew up in a household where my parents grew food, and the honest answer is that I did, but not as most people would imagine it. It wasn't a positive influence – quite the opposite, actually. My mum came from a farming family, and she knew how to grow food naturally. Her family grew food and raised livestock overseas, they lived sustainably and were on top of their game. Unfortunately, my father, who was a city person, learnt to garden here in Australia, and always dominated the whole backyard garden with mass plantings of a single crop, usually tomatoes, leaving only tiny pockets of random space here and there for my mum to grow flowers. My younger brother and I would have to slavishly stand at the tap for what seemed like eternity, waiting in the hot summer sun for the command to turn the tap on or off as my father moved the hose from garden bed to garden bed and flood-irrigated the soil to the point of turning it into a swamp. He also had an obsession with toxic pesticides!

Memories of 'tomato dust' covering the leaves of plants like icing on a cake are best forgotten. My father was obsessive with his mini-farm method of toxic monoculture production that emulated the common practices of modern agriculture. At least we got to stay inside when he put on a towel around his face as a makeshift respirator and doused the garden – and inadvertently himself – with highly poisonous pesticide sprays. It really was an experience to put someone off gardening for life, but fortunately I prevailed. I had a gut feeling back then that this was not the way food was supposed to be grown. Later in life I would learn that my suspicions were indeed correct.

My first real experience with plants was in primary school, and it was a positive experience, which sowed the seeds that many decades later would determine my future career path. During my second year of primary school, we were each given a small catnip plant in a pot, and over the coming weeks in class we cared for and tended to our plant and observed how it grew. The novelty factor with catnip is that it's a cat-attracting herb, and for kids that was pretty exciting. Eventually, our plants grew big enough that we got to take them home. I asked my mum to help me plant it in the garden, and that was the last of it, or so I thought.

I would describe my childhood education as highly irregular. Unlike other kids, I grew up reading and understanding secondary school biology and chemistry textbooks in primary school, thanks to my dad's misguided but well-meaning efforts. It was no surprise that I went on to study science at Monash University, with a double major in the biomedical sciences, pharmacology and toxicology, and biochemistry.

A DETOUR ON MY LIFE'S PATH

Through sheer chance, for my Year 10 work experience appointment, I found myself in the Victorian Forestry Commission. I was only meant to be there for two weeks, but because the scientists saw my enthusiasm and realised how much I knew about laboratory work, when the two weeks were up they invited me to work there permanently as a lab assistant rather than go back to school. That would have been an instant career in forestry research, but I had already made the choice to continue school and go to university to study science.

My university years flew by, and I graduated in 1989 only to find that were no jobs in science. It figured that I'd studied science for love, not money. I deferred my honours year then decided against postgraduate study as the job prospects were not good, even with higher qualifications. The world of business, by comparison, was booming with jobs.

So, instead, I went to a technical college, studied bookkeeping, and then entered the corporate world doing office administration work for the Melbourne Fire Brigades Board. From there I got a job with the Department of Housing, working as an accounts payable clerk, and in a few years worked my way up to being an acting finance manager. All my life I had been a quiet-natured problem solver and now, as a 27-year-old manager, I was responsible for a team of people. Luckily, I had the support of a visionary manager who was keen on developing my skills. This was a pivotal shift in my life from being focused on things to focusing on people, a shift that would ever so subtly allude to my future work. I learnt to function like an extroverted 'people person', albeit for a short time only.

I was eventually promoted to a financial analyst role and once again immersed in the impersonal world of things. Not long after, I commenced a postgraduate degree in accounting. I only completed half of it, as I found it quite boring, but in the process I learnt enough to work as a management accountant. In that job I taught myself computer programming and singlehandedly designed and built the department's financial-management reporting system. I had spent about seven years in the field of finance, and it seemed like time for a change. I felt unsettled, like a wandering spirit seeking a world that I could call home.

Having previously worked with computer systems, I moved into information technology, a field to which I could readily transfer my research and analytical problem-solving skills. I undertook even more study and got my industry certification as a systems engineer, and from there I rode the crest of the wave of rapidly growing IT industry. As a systems engineer, I worked in the banking, finance and then the insurance industries. Eventually, I moved to contracting work, which paid more. I ended up staying in the IT industry for ten years, which was a long time to spend working on impersonal things such as computers.

After working for roughly 20 years, I realised that I had well and truly strayed from my interests in the natural world. I was at a crossroads in my life and had no idea which was the right road to choose, but I was never required to make that decision in the end. My father was a lifelong chain-smoker, and was slowly smoking himself to death. As his health declined he turned to herbal medicines, and on one fateful day my mum asked me if I could grow some herbs for him. I agreed.

One simple question and an affirmative reply was all it took to change my life and potentially that of many others worldwide.

HERBAL FIRST STEPS

As I ventured off to Lilydale Herb Farm to purchase my first plants, I reassured myself that I wasn't starting gardening completely from scratch. I knew something about medicinal herbs, as I had grown up reading and re-reading my father's tattered copy of the classic in Western herbalism, Nicholas Culpeper's *Complete Herbal,* first published in 1653. I figured I knew a tiny bit, having studied some botany at university. I was a very practical person and was good with my hands. Optimistically, I looked forward to the opportunity of learning and gaining new skills, and since gardening was meant to be fun, what did I have to lose?

For a beginner learning the ropes, my approach was to read, then do. I would learn as much as I could about the task at hand, and, when I felt confident enough, I would step out into the garden and get my hands dirty. I decided to grow all my herbs in large pots because I felt that container gardening was less intimidating than managing a large garden bed. My early efforts were rewarding, my herbs grew well, my father had a constant supply of medicine and my mum saved a fair bit of money. In the process I discovered that I really enjoyed gardening.

In a very short time I had amassed a collection of around 100 pots of herbs and other interesting plants. Summertime, when plants are most prone to wilting, is a real wake-up call that gardening is about caring for another living thing. It was clear that these plants I had adopted were dependent on me, because I had artificially restricted their capacity to extend their roots for water and nutrients, by putting them in pots. It was now my duty to keep these plants alive.

When you start caring for living things you really start learning about life. Bill Mollison, the father of the Permaculture movement, once stated that 'gardening can teach you philosophy, but philosophy can't teach you gardening'. I reckon he was right! Through interacting with nature while gardening, I learnt it took commitment, dedication and care to nurture life, while destruction of life took very little effort and only a brief moment of thoughtlessness.

For me, working with nature also felt like a small act of gratitude. When we stop long enough to notice the subtle undercurrents of life, we become acutely aware of our complete dependency on nature, how we are sustained through the air, water and food we need to survive. The act of giving a little something back felt like a way of acknowledging what I was receiving. The more time I spent with plants, the more I was amazed by them. I had finally come to see what the early scientists saw, the richness, complexity and sheer beauty of nature. What they saw inspired in them a true reverence for nature, and it had the same effect on me.

GROWING ORGANIC FOOD

After learning how to successfully grow herbs, I began looking for my next gardening challenge. Recognising the value of natural, fresh healthy food, I decided it would be great to try growing my own. Naturally, I went straight to using organic gardening methods for food production, strictly no chemicals or poisons. After all, as a qualified toxicologist, I know all too well how bad pesticides and herbicides are for all living things and the environment, how the poisons applied to food crops ultimately end up in our bodies and what harm they cause us in both the long and short term. Poisoning may have been all the rage in medieval

Europe as a way of dispatching adversaries but, frankly, I prefer my food unpoisoned, and know better than to trust the hype and misinformation coming from corporate marketing departments.

My journey into organic gardening began in 2002, and I learnt a lot about gardening sustainably and minimising my environmental impact. I soon came to realise the delight in growing one's own food. You know exactly what's gone into it, you become familiar with the habits and nature of the plant, and you in fact *commune* with nature. It's a relationship, an exchange, and, best of all, it's all for free. Compare that with the detached monetary transaction of purchasing some poor violated plant product from a supermarket without a second thought. The experiences really are worlds apart.

One other thing that growing food did was to make me think about our society's relationship to food. All food comes from nature, and, as we are actually *made* out of the food we eat, that's a fairly intimate exchange – even more so when we're connected to the process of growing the food itself in some way, directly or indirectly. I realised that when we become disconnected from our food, we don't treat our food with proper respect as a gift from nature, and, as a consequence, we lose our connection with nature and our proper sense of place in the world ecologically.

Beyond philosophising, I eat just like everyone else and enjoy the taste of good food, so I set out to understand what makes food taste good. I wondered why supermarket produce tasted awful and was of such a low quality. Everyone who has ever smelt or tasted an organically grown apple can vouch for the amazing aroma and flavour of fresh natural produce. It's entirely different from the small, hard, scentless, uniformly round and unnaturally wax-polished 'product' impersonating a real apple that you buy from a supermarket.

I looked into how people could make natural food this bad, and found that it's no big secret – perhaps it is to the general public, but not to the agribusiness sector, who incidentally don't see any need to hide it. It's purely pragmatic in their minds. To be able to pile fruit in huge quantities in a big truck and ship it vast distances without bruising, it's picked unripe and rock hard! This means that the flavour, sweetness and optimum nutrient levels aren't allowed to develop naturally in their own time. These horrible little unripe fruits then go into cool storage for up to *a year,* during which time they are treated with a gas that suspends their metabolic processes. When it's time to go to market, they are put into a chamber which is filled with ethylene gas to artificially ripen them. This doesn't work for lemons – they stay green – and so the industry uses yet another chemical to get the yellow colour. Often the apples and many other fruit are shipped to another country to be waxed and polished with petroleum-based paraffin waxes, which may be blended with vegetable-based waxes, and which may include dyes for cosmetic appearance and also fungicides, then shipped back for sale to consumers who expect fruit to be 'in season' all year round. Bon appétit!

In project management there is a concept of a 'project triangle', where you have quality, cost and time – pick any two, because you can't have all three. It's the same for all retail products. If we apply this concept to food, we have the following choices: we can have something that tastes good and is packed with nutrients, something that is cheap, and something that is available immediately. The public desire for fruit on demand means that they choose 'time', and wanting cheap prices means that they choose 'cost', so something has to give, and that's quality. I wasn't happy to sacrifice quality, or to have this choice made for me by a supermarket duopoly. My choice to grow my own food, and support farmers who grew quality food, seemed logical and rational.

Digging even deeper, I also discovered that the lack of taste was a subtle indicator that something else was missing, and that something was nutrients, the primary reason why we eat food. I understood that plants can only contain nutrients they can find in the soil, and it stood to reason that since agribusiness grows its product in lifeless, depleted soils, the nutrients would be lacking. The way they get plants growing in bad soils is to force-feed them with chemical fertilisers, which contain the bare minimum nutrients and minerals to make the plants grow. It's about cutting every corner possible to maximise profits – in Australia it's the supermarket duopoly that reaps all the profits, the farmers just try to cut their costs any way they can to stay alive. When I considered these practices of modern agriculture, I was not surprised that conventionally grown food had a lower nutrient density. What consumers don't realise is that they get fewer nutrients for their dollar, meaning they have to eat more of this lower quality food to get the same amount of nutrition.

Our taste buds can tell us only so much, so when I checked the research on nutrient density, I found that in 2004 the *Journal of the American College of Nutrition* published a paper by Donald R Davis titled 'Changes in USDA Food Composition Data for 43 Garden Crops, 1950 to 1999'. The research showed a decrease in the levels of calcium, iron, phosphorus, protein, riboflavin, vitamins A and C of 5–35 per cent in conventionally grown produce. More recently, in March 2008, Charles Benbrook published research that showed that organic foods were much higher in antioxidants, nitrates, polyphenols and vitamin C, and slightly higher in protein.

I figured if I was going to be growing food, I would want to be growing the tastiest, healthiest and highest quality food possible – why would I opt for anything less? I kept growing food and continued to read, research and learn. I experimented with more sustainable gardening methods for five years, and at that point I thought I'd seen it all, until one unexpected turn of events.

DISCOVERING PERMACULTURE

Through sheer serendipity on the internet, I stumbled upon a video that totally captivated me. It featured an old Australian larrikin with a white beard, lying in a hammock, surrounded by food growing all around and above him, looking as comfortable as can be. I remember him reaching up and casually picking some fruit dangling overhead, eating a mouthful, then looking at the camera and uttering something to the effect of, 'This is the life!' This man was talking about the importance of growing food – the scientific, ecological and social reasons. He talked about the importance of soil health. He clearly knew what he was talking about, and I needed to find out more about this enigmatic figure, so down the rabbit hole I went.

This character was none other than Bill Mollison, a Tasmanian ecologist who travelled the world and taught communities how to grow food, and who, together with David Holmgren, founded the applied science of Permaculture in the 1970s.

What struck me as revolutionary about Permaculture was that it proposed that we could take a fully sustainable natural system and work harmoniously with it to supply our needs. This made more sense to me than what modern agriculture was trying to do

– reinvent the wheel. Had nature not already evolved the best way to grow plants sustainably over the last few hundred million years? Did nature not run fine on its own for 95 per cent of its existence before the first humans appeared on the planet?

What is Permaculture?

If someone asked me today, I would tell them that Permaculture:

- is a holistic design system
- emulates systems that exist in nature
- creates sustainable human settlements and food production systems
- integrates harmoniously with the natural environment.

The US National Sustainable Agriculture Information Service has defined it as:

- ecological engineering
- cultivated ecology.

The word Permaculture is a portmanteau of the words *permanent agriculture* or *permanent culture,* which alludes to the goals of this system.

This made me critically look at what agriculture was trying to do: namely, trying to grow plants in a way they did not grow naturally, in places where they would not grow naturally. Vast amounts of energy and resources were being poured into the never-ending task of killing the plants that nature wanted to live, while propping up the plants that would naturally die. Even worse, agriculture was actually fighting against nature, resorting to full-scale chemical warfare in the process, and burning fossil fuels at an alarming rate to combat nature. The idea of using the planet's finite fossil fuel reserves to push against nature's processes, which are powered by the sun, which for all intents and purposes is inexhaustible, seemed incomprehensible.

As I delved deeper into Permaculture, much to my surprise, I found it went a step further than other sciences, which are essentially amoral in nature, by introducing ethics into the equation. The three ethical principles of Permaculture are simple but far-reaching: care of the earth, care of people and return of surplus to the earth and people (also called 'fair share').

I looked further into Bill Mollison's work and came to realise he was truly a visionary ahead of his time. He was warning the world about the issues of food security, climate change and soil erosion back in the 1970s. Society is only starting to take action now, 40 years later. Bill's work was widely recognised around the world, and many communities worldwide were thriving after adopting his systems, but Australians were largely ambivalent. There's a 2000-year-old saying that 'a prophet hath no honour in his own country', and it looks like that adage still holds true.

In 2007, I embarked on a year of self-study, working though the book *Permaculture One: A Perennial Agriculture for Human Settlements* by Bill Mollison and David Holmgren. This book was filled with inspiring ideas, and was one of the factors that prompted me to build my first Permaculture garden. I had some time off between jobs while contracting in the IT industry, and over a three-month period, I methodically researched, designed and built my garden, one garden bed at a time. I finished my garden in late 2008, and was content with having taken my organic garden to the next level in sustainability. I had come a long way as a self-taught gardener since 2002. This point may have been a comfortable place to stop, but that's not what the future had planned for me at all!

CONNECTING WITH OTHERS

Life feels like synchronised clockwork when you just happen to be in the right place at the right time, and what you need lands right in front of your feet, even if you weren't aware that you actually needed that something. That was precisely the experience that led me to formally study Permaculture. I'd only just finished building my garden two weeks earlier and, while reading online about Permaculture, I unexpectedly stumbled upon an advertisement for a Permaculture Design Certificate being run in a few weeks' time at Trinity College in Melbourne. Much to my surprise it was being taught by Bill Mollison himself, along with one of the world's leading Permaculture experts, Geoff Lawton, and a new teacher, Greg Knibbs, from Western Australia. I'd never imagined that the man in the video, the co-founder of Permaculture, could be my teacher. I deliberated for only a moment, and then I thought to myself, 'Stuff it, what have I got to lose, I'm enrolling!'

I went in expecting to learn more about Permaculture. What I didn't expect was that this 80-hour intensive course would be a life-changing experience for me, and for all the other students who flew in from around the world to be there. Through Bill's traditional storytelling to convey timeless wisdom from many cultures, Geoff's highly precise and methodical explanation of technical facts that you would expect from someone with an engineering background and Greg's intimate understanding of human communities, we were taken on an experiential journey of learning that felt more like a deeply profound initiation.

We all spoke about the experience after class, and we even came up with a name for it: 'the Permaculture effect'. Initially, it was hard to pin down what had happened to us all. You couldn't find a more diverse group of people – there were scientists, engineers, farmers, IT people and even artists in the group, and they all came from different countries and cultures – yet everyone felt the effects equally. We finally figured it out: our perspectives had simultaneously shifted and broadened, and on some deep level we finally understood where we, as people, fit in on this planet. This wasn't a mere intellectual understanding or state of opinion either, it was a completely unfamiliar way of looking at what has always been there before us, something the ancients described as 'seeing the world through new eyes'. Words can only at best approximate the experience and in no way do it justice. Something had changed irreversibly, and there was no going back.

If that experience wasn't enough, there was still more self-discovery, for me at least. Towards the end of the course, each student had to stand before the audience of more than 60 fellow students and give a speech. I wasn't afraid of public speaking, as I was already familiar with attention from the newspaper and television crews in my younger years, but I wasn't sure how well I could pull it off. Not having much time to plan, I decided hastily that I would talk about what I had learnt, and what I intended to do in the future, which was to leave the world of IT and dedicate my life to something that made a real difference to the people and the planet, something ethical, something the Buddhists call 'right livelihood'. I didn't realise how impassioned my speech was until the hearty applause at the end and the comments from my fellow students that my talk was really inspiring. I questioned myself: could I talk and inspire people? I didn't realise at the time this was a sign of what was yet to come.

On graduation, I felt so lucky to be a student of such a great man as Bill Mollison, a true Australian pioneer. He received the Right Livelihood Award in 1981 with Patrick van Rensburg, a South African educationalist and former anti-apartheid activist. This was a man who really made a difference in people's lives. As I walked up to shake his hand and receive my certificate, I said to him, 'One day I want to be great, just like you,' to which Bill smiled and replied, 'I'm just an ordinary bloke.' Without time to think, I just spoke from the heart and calmly stated with firm conviction, 'Then I want to be ordinary, *just like you.'* He grinned because he knew what I meant, that I wanted to make a difference.

Inspired to apply what I'd learnt, I made more changes to my garden. Geoff Lawton had taught us about food forests, the most sustainable garden system possible. He explained that food forests are basically modelled on a temperate forest, with multiple layers and levels of plants, but use edible and other useful varieties of trees and plants instead. Most importantly, he emphasised the fact that food forests have one major qualitative difference from any other garden or food production system: they are living ecosystems, and function as such. As a result, natural processes take over to regulate the system as they do in nature, and when this happens all the beneficial insects, animals and birds move in – and take care of all the garden pests, naturally and for free! Just like a real forest, it was a system that did not need any spraying, weeding or even digging, just watering, some pruning and lots of harvesting of fresh, natural food.

This all sounded too good to be true, but I didn't doubt Geoff's technical competency, as he was a technician par excellence – he had a background in engineering, a technically meticulous approach and an impressive portfolio of pioneering work. A master of earthworks and hydrology, he had restored dry farmland and regenerated natural springs, but his most impressive work was the Greening the Desert project in Jordan, where he turned an expanse of dry sandy desert into a lush, abundant food production site, documented for everyone to see. A basic principle of science is that all sound science is *reproducible,* so I decided to test the theory for myself and embark on a huge experiment, to build a real food forest in my backyard.

The other factor that prompted me to build a food forest was the lack of documented examples and case studies worldwide. I could find examples of rural food forests, but no urban ones. Of the ones I could find, the information was limited, with no hard figures for yields and no detailed documentation. I found no shortage of uninformed opinions permeating all manner of online discussions though. There were people claiming Permaculture didn't work at all, or that it only worked on small sites, while others insisted it only worked on large sites – all opinion with no evidence to back up the claims. I figured it was time to silence the critics with a fully documented case study, which would be published online.

MY FIRST FOOD FOREST

Doing something the first time round is never easy. Doing something that potentially has never been done before takes a lot of creative thinking and a whole lot of patience.

My project was definitely ambitious, and my first technical hitch was to work out how I would fit a food forest with many fruit trees into an 80-square-metre backyard garden. All documented food forests were built in rural areas and used large trees. Some were nothing more than mislabelled orchards, which are clearly nothing like a food forest. As far as I could ascertain, no one had ever built and documented a real urban food forest in a backyard. My research on small-scale orchards led me to a technique from the United States called Backyard Orchard Culture. In this innovative method, which uses summer pruning as opposed to traditional winter pruning, full-sized fruit trees could be dwarfed to any height without using special dwarfing rootstocks, which incidentally are more expensive, slower to grow and aren't available for all fruit trees.

My pioneering work was to combine this scaling down of trees with the multi-layered system of food forest gardening, a system that can comprise up to seven layers – large trees, with small trees underneath them, shrubs, herbaceous plants, ground covers, climbers and root crops.

As a corporate workplace trainer and technical writer, I enjoyed writing instructional training material on a whole host of sustainable food production topics on my website, *Deep Green Permaculture,* and traffic started increasing steadily. For many years I ran my website under a pseudonym, because my aim was to draw attention to my message rather than to myself. It mattered little who I actually was, and only when people figured out I was behind the website did I finally put my name to it. As I progressively documented how I did everything, *Deep Green Permaculture* gradually gathered a worldwide audience.

Recording the yields of my garden over the first four years silenced the critics permanently. I had squeezed in 30 young fruit trees, a dozen varieties of berries and over 60 medicinal herbs, and in the fourth year, my pioneer urban food forest garden produced over 230 kilograms of food – 160 of fruit, 60 of vegetables and 11 of berries.

What is astounding is that this is the equivalent of around 14 metric tonnes per acre, while an Australian dry-land wheat farm produces about two tonnes per acre, and the best European farms can get around four to eight tonnes per acre, and that's using chemical fertilisers and toxic pesticides and herbicides. My food forest garden is naturally pest-free and has been that way for years. It's a water-wise garden that runs purely on rainwater for half the year, and with two 50-minute waterings a week the other half (and some supplementary watering during extreme heatwaves), and can be maintained with an average of two hours' work a week.

These results very quickly got people's attention, and the truth is that the figures are actually *understated!* Most years the garden effortlessly produces 70 to 75 kilograms of vegetables, and in a garden full of medicinal and tea herbs, herb harvests are not included in the yield figures. The garden is now capable of far greater production as the fruit trees are more established, more trees have been added, and the amount of berries has almost doubled. Being a decade-long experiment, the real potential is yet to be realised. I decided to take a short break from recording produce after year four, with the intention of resuming in years seven to ten of this study.

Slowly, I started seeing the results of my work. After six years of advocacy work, educational talks and demonstrations, people began talking about and even building food forests. I ran garden tours and the idea of Permaculture started seeping into the mainstream. I figured that I was able to reach the mainstream because of my approach. I wasn't preaching an ideology, nor to the converted. Reasoning that the majority of people live in cities, and the majority of those people are regular people, I logically directed my efforts towards the general public, where I could reach the widest audience. Despite what our backgrounds are, one thing that all people have in common is food!

I was able to show people a real-life example of what was possible, and, most importantly, I explained how they could do it themselves. That was my message, and only after they understood it all did I finally mention that it's called Permaculture. Over the years, thousands of Permaculture students, community groups and members of the general public have visited my garden and found it inspiring. My garden won the Darebin Council Sustainability Award in the House and Garden category in 2013, and I was invited to take part in the prestigious Australian Open Garden Scheme that year, where my garden received the most visitors of all the produce gardens in Victoria, with 480 people visiting over a weekend. In 2014, I was also given two runner-up awards for community leadership in sustainability by the City of Darebin for my work.

I figured I was on the right track in the very beginning when the Permaculture Research Institute sent down a photojournalist from interstate to do a photo shoot and write an article on my demonstration food forest garden. A few years later I knew I had achieved something significant when my teacher Geoff Lawton flew down from interstate with a film crew to record footage of my garden as part of an urban Permaculture video he was filming. The *Deep Green Permaculture* website grew to become an international resource for sustainable food production, with over 2000 views per day across 120 different countries. I think I have achieved what I intended, to do something that mattered. Or at least, I have started!

BROADENING FOCUS TO THE COMMUNITY

Just as a snake sheds its skin to grow, my experiences had changed me as a person, and I had outgrown my former life, so I decided to leave the IT industry forever. I entered a whole new world, filled with people who held values similar to my own, who dared to dream of a better world, who were driven by an inner stirring to make a difference. I spent many years working with my newfound community, doing voluntary work, sharing my knowledge, establishing my reputation. When the time came, I began a new career as a trainer and presenter in the area of sustainable food production.

As my work became more well known, local governments became interested and invited me to teach the general public about Permaculture and how to grow food sustainably. This to me was a significant paradigm shift, as Permaculture has had very limited success in reaching the mainstream. As people saw my food forest garden, it inspired them. Word of mouth spread and more people came. Permaculture teachers brought their students to see a working food forest system in an urban area. Not only had I put food forests back on the map in the Permaculture community, but I broke the concept out into the general community, and made it accessible to them. I never planned this, it just happened as I followed my heart, and I eventually became a food forest specialist.

I was happy with my progress, but I still felt something was missing. What had I really achieved? I had built a demonstration site in my own backyard and proven how much food it was technically possible to produce with limited space, showing the world what one person can achieve on their own. That last part was where the problem lay. Humans are a social animal, and communities are cooperative. As the saying goes, no man is an island.

It was time to expand my successful technical model to include a critical element – community. I realised that to feed communities we would need a collaborative model of food production. Perhaps we could create many small food forests right through the urban landscape, and link them via a collaborative community where all produce was shared. Maybe we could have food forests on a larger scale than was possible in backyards, where people could work collaboratively together. There were many possibilities, many ideas, but ultimately when it comes to feeding people, trees and plants need to be planted in the ground. I knew it was time to take the next step, to extend the food forest garden into public space, and I figured a community garden would be the ideal site.

Shifting to a focus on communities meant that my work had now become more about *urban agriculture.* For some, the idea of people in cities growing food is a radical concept, but when seen in a proper historical perspective, we realise it's simply the act of people reclaiming their sovereign right to produce food, once again restoring the ecological status quo of living in close association with their food, just like all other living things have always done. My new objective was clear: to fill urban areas with sustainable food, and to convince others to do so too.

Why fill cities with food? Because food is clearly absent from cities, and food is one of our basic needs. If our basic needs aren't met in abundance, how can we define ourselves as prosperous? A city brimming with food growing everywhere, where everyone is housed, is a sign of real prosperity in my mind. Measuring our prosperity by the number of pointless electronic gadgets per capita is ultimately quite misdirected. When I investigated how we ended up with food-free cities, what I found was a classic case of people doing things simply because that's what people before them did – not knowing the reason why, and never asking the question. The original reasons become long forgotten, but the inappropriate and often pointless actions continue.

Our food-free cities started back in seventeenth-century England, with the deliberate growing of trimmed grass by the wealthy and the aristocracy as a show of affluence. During those times, land was a valuable resource used for growing food, as a source of nutrition and income. For a wealthy landowner to simply grow grass was an extravagance. These lawns were cut by hand by servants using scythes, sickles and shears. The labour involved in maintaining a large lawn was considerable, so only the wealthiest in society could afford to pay people to carry out this work instead of growing food for them. The next level of decadence came with the introduction of formal ornamental gardens, when the trimmed grass was complemented by meticulously manicured ornamental gardens, for aesthetic purposes only.

As bad luck would have it, the influences of the English and French formal gardens, originating in countries with plenty of water and milder weather, carried into the new vaguely defined profession of landscape architecture, even in countries with little water and intense heat such as Australia. The cultural inappropriateness should be glaringly obvious, and the inherent wastefulness should be highly questionable, but, nevertheless, the landscape architects were employed to 'green the cities', often working for government, carrying on the gardening-as-art tradition. Culturally, we were emulating the people who were emulating the people who were once wealthy, having lost sight of why it was done. Additionally, there was a cultural stigma around growing food – that's what poor people did!

The southern European migrants who first migrated to Australia in the 1950s were laughed at for growing produce in their front gardens. They came from a culture that valued food, and many of them were farmers, who were truly self-sufficient. They were dismayed at the extremely limited number of vegetables here, so they brought seeds from their countries and grew their own vegetables. Having spoken to many first-generation migrants, my own parents included, they were always puzzled why all the Anglo-Celtic people in Australia only grew 'useless plants' and nothing edible. Ironically, many Australians with an environmental focus and a desire to become self-reliant, typically in the 20- to 30-something age bracket, and a few of the older generation, are now growing produce in their front yards.

The high aspirations to produce all of one's own food and live sustainably seems like a big deal to many Australians, especially to those of Anglo-Celtic origins, but to those 'old wogs' who bore the brunt of racist derision for their food-growing practices, living that way was always second nature. To people here from other cultures, our situation appears bizarre. A friend's husband who came from overseas about a decade ago expressed his concerns about Australia's cities. Where he came from, he would go jogging and pick fruit along the way to eat to keep his energy up or snack on if he got hungry. Food growing in public space is commonplace in his country, Cyprus.

What interested me was how such cultural biases managed to be maintained in this day and age, so I asked landscape architects why they didn't include fruit trees in public landscapes. I got the nonsensical and often repeated argument about messy fruit falling on the ground and public liability if someone slipped. I went looking, and what I did find was that any fruit trees overhanging into public space were picked clean by the general public, often unripe. Further investigation revealed that there were even 'fruit squads', organised groups that mapped locations of and harvested publicly accessible fruit, which they then donated to charities that fed people.

When I looked deeper into the education of landscape architects at graduate and postgraduate level, I found out they barely touch on the subject of trees and are actively discouraged from getting too caught up in horticulture. The discipline is meant to be a high-level conceptual one: they are trained to drag round shapes that represent trees on their computer screens in computer-aided design programs, and apart from learning about a few basic ornamental tree varieties, that's as far as it goes. When I asked one landscape architect, he honestly explained that in the workplace they have tight deadlines so they just use the standard ornamental trees that everyone else uses.

The reasons why we don't grow food in cities are misguided at best. But there are many good reasons to grow food in cities and promote urban agriculture, and one of those very good reasons is food security.

This became apparent to me one day on a bus ride home, when I contemplated where people would get their food from if society collapsed. On the circuitous half-hour bus trip, I watched out the window to see if there was anything edible growing, and much to my amazement, I saw nothing. Supermarkets had a limited supply of food if the supply chain was somehow disrupted, so this was really a badly designed system with many points of failure and no back-up. A complete dependency on rural agriculture and fossil fuels – talk about putting all your eggs in one basket!

Permaculture has a design principle that states, 'Each important function is supported by many elements.' The reason for ensuring that critical functions, such as food production, are supported in more than one way is essentially that of *resilience.* A resilient design is one where the operation of critical functions continues if any one system breaks down. If we grow food in our cities (urban agriculture), our city fringes (peri-urban agriculture) and in the remote regions (rural agriculture) we can create a very stable, resilient food production system.

I was disappointed to find that our esteemed horticultural colleges didn't even consider urban agriculture a subject worthy of being taught. I also found that urban agriculture was belittled by the rural sector, the attitude being that it was a token gesture, but some of the industry was clearly threatened by the growing movement of urban food production. I've been surprised by people who are willing to criticise urban agriculture without any real facts behind them. During one radio interview on a program aimed at rural agriculture, I upset the host by correcting him on his assertions about urban agriculture when I explained that the predominant state of affairs throughout human history has been local agriculture, and the phenomenon of rural agribusiness has only existed in the last 200 years.

When he challenged me on the issue of productivity, I explained how many small gardens spread across the country create a vast decentralised system that is highly resilient, and then I did something he got very upset about. I gave him hard facts and figures that showed that US home gardeners, with small gardens averaging around 600 square feet (55 metres), collectively grow more food than the whole Australian agricultural industry put together.

Even though most Australians are considered 'food secure' and the Australian federal government brags that we export two-thirds of the food we produce, it's ironic that state governments fuss about water shortages, considering that agriculture uses around 80 per cent of all the nation's fresh water. In reality we're exporting our most limited resource, water.

Furthermore, the food charity organisation Foodbank, citing the findings of an ANU survey conducted in 2012, reveals that '4 per cent of Australians are accessing emergency food supplies and, in total, 8 per cent say they are unable to afford food. Up to 16 per cent of respondents say they often or sometimes worry that their food will run out before they have enough money to buy more, and 13 per cent say they cannot afford to eat nutritionally balanced meals.' According to their estimates, around 2 million Australians use food relief each year. In a report for the prime minister's Science, Engineering and Innovation Council, we are told the facts:

> The likelihood of a food crisis directly affecting the Australian population may appear remote given that we have enjoyed cheap, safe and high-quality food for many decades and we produce enough food today to feed 60 million people. However, if our population grows to 35–40 million and climate change constrains food production, we can expect to see years where we will import more food than we export. We are now facing a complex array of intersecting challenges which threaten the stability of our food production, consumption and trade.

Unfortunately, the government's response to food security as seen in the National Food Plan was rather weak-willed and misdirected, focused primarily on finding solutions through big agribusiness, and looking to fund more research into the industry's holy grail of genetically modified crops to deliver bigger yields while carrying on with a business-as-usual approach.

Bill Mollison taught me that if we wait for government, scientists and big organisations to do something about the environment or food security, nothing will happen. Timely and responsive action to initiate change comes from the grassroots level, from individuals, collectives, communities – this is where social change starts. Collaborations with government can work, but expecting the 'authorities' to fix things is a recipe for failure. One idea that came strongly out of my study of Permaculture was the notion of self-responsibility, responsibility for my community, and responsibility for the planet. It was soon quite clear in my mind that advocacy work was important, and I began the work of shifting perspectives.

A COMMUNITY FOOD FOREST

Through teaching in my local community, it wasn't long before I met people who worked with me to take my work to new heights, and who invited me to work with them so that it became *our* work, a creative synergy that would see us all reach new heights. The great thing about being a presenter is that you get to meet so many interesting people during events, and when people who are guided by a common vision and set of values come together, amazing things happen.

During one talk I met a woman named Charlie Si, a very active networker who links people with causes, events and groups. She immediately saw my potential and knew how to bring it out by connecting me to the right people and events. I shared my vision with Charlie about building a food forest in public space, and in no time she was back with good news: the Northcote Library Food Garden was a community garden that consisted of a series of empty garden beds, and the council was looking for a community group to come together to build it. It was our big chance. Charlie brought in advisors – of which I was one of four – as well as gardeners and a management committee. There was no funding, so she also gained the support of other community gardens, who donated plants, and the whole site was planted up in a single day with plants that were given to us. The community garden employed a unique collaborative or shared model where everyone worked on the garden together, and everyone shared the produce, and it worked well.

Finally, I had built a food forest in public space, a food forest that the general public could walk through, pick and eat. It was scaled down for children, and employed a multi-sensory garden design that children could explore. It was delightful watching parents bringing their children to forage for alpine strawberries while the fruit trees were still growing.

The Northcote Library Food Garden was completed in 2011, three months after it was first constructed, and it became a hub for community activities, training, festival events and talks. It successfully incorporated the elements of food and community together, reconnecting people with food and each other. Inspired by this success, Charlie Si, John Pinniger – the co-founder of both Permaculture Victoria and the Heritage Fruit Society – and I decided to run an event at the Sustainable Living Festival to rally support for an even bigger food forest, in wide-open public space at All Nations Park in Northcote, a reclaimed landfill site that was transformed into a huge public park. We ran a tour for a group of local people and some council representatives through the park and shared ideas of what was possible. It was a walking brainstorming session, lots of great ideas came out of it, and the council saw what public support existed for the idea.

After a lot of research into the history of the site, another colleague and I proposed the idea to our local Greens councillor, who promised to run with it. The proposal was passed through council and approved as a local government project. Eventually, we got word that the council had received funding for the project, and we were a step closer to realising our goal, the construction of the All Nations Food Forest. What was special about this project was that it would be a real first; no council had ever built a food forest. Some food forests had been built by the general public on land lent to them by councils, with very little support or funding, but this was different; it was a big-budget project in open public space, with proper soil amendment carried out using earthmoving equipment and even professional irrigation installed. As a demonstration site, it would change the paradigm of food production from what people were familiar with to a more ecologically sound and sustainable method.

John Pinniger and I acted as technical advisors working with the council's sustainability team and landscape architects. The council engaged in extensive public consultation and ran events where the public were given input about the type of food they wanted grown. I collated the feedback, reviewed it from a technical perspective, then made my recommendations for the varieties of fruit trees suitable for a low-maintenance, open-space public site, and these formed the horticultural foundations of the project.

The food forest was completed in late 2014. It was built as a terraced raised-bed design on the sloped site as per my recommendations, and the extensive hard landscaping, as designed by the council's Open Spaces team, was constructed by contractors in a few weeks. Once the construction work was completed, the council advertised an event where the general public would plant approximately 30 fruit trees and hundreds of understorey companion plants. It was designed to be an educational event. My role was to demonstrate how to plant trees and plants, and the people attending the event would learn through hands-on practice. We stated in the advertisements that the event would run for four hours, but so many people came that the whole site was planted out completely in only 90 minutes!

As we wait for the garden to establish itself, we're busy establishing a community around All Nations Food Forest. We've already run a pruning workshop and the first working bee for the general public. It is envisaged that this will become a site where people can gain more skills in food production, learn basic gardening skills, and share their knowledge and experience as well as their love of good food. It is a site that promises the benefits of community building, public education and food security. The intention is also to use it as a template for future projects elsewhere, a precedent for other councils to follow. Councils are more likely to attempt similar projects on this scale and level of ambition in the future when they see that another council has done it successfully. Building a food production site on converted landfill with potential soil-contamination issues is the most technically challenging location possible, so if urban agriculture can work here, it can work anywhere. One important thing this site demonstrates is that almost any space can be used to grow food.

I'm a strong advocate of urban agriculture demonstration sites. People trust their senses and what they can see with their own eyes. More so, they can taste what's growing there. My goal was to put food back into cities, and I've managed to do this through a backyard demonstration garden, a community garden and an open-space public garden. I am also affiliated with an extensive network of like-minded people, with whom I often collaborate as part of an urban agriculture grassroots movement that is busy building food gardens in backyards, community spaces, schools, parks and wherever space makes itself available. Urban agriculture is definitely on the rise in cities as people reclaim their sovereign right to grow food.

I've decided there are more boundaries to push, and I'm now building multi-sensory food gardens that appeal to all senses – they don't just taste good, they look good and smell nice too. Marrying my biomedical and horticultural knowledge ideally positions me to tell the world about wellbeing gardening, gardening that benefits the body, mind and spirit from a scientific evidence-based perspective. Then there's the concept of food as medicine, an idea drawn from my studies of Eastern culture. Where does it end? It doesn't, it's a life-long journey.

Incidentally, there's a primary school project that I've been working on with my colleague Kat Veilgaard Irwin to introduce children to growing and eating fresh food. She has done a lot of work with children's gardens and I leapt at the opportunity to work with her because I know firsthand how even a brief introduction to gardening can change a child's life.

I still have the catnip plant I brought home from primary school almost four decades ago. It's growing abundantly in my food forest garden. That plant found its true home in a food forest, as have I in food forest gardening. Once I finish working on that primary school food garden, maybe I'll grow a bunch of small catnip plants and give them to the kids. You never know where it might lead...

FINDING A LIFE THAT MATTERS

KIRSTEN LARSEN

Kirsten Larsen is one of the leading thinkers and, increasingly, 'doers' in this field in Australia, and she has been for many years, through her work with the Victorian government, the Victorian Eco-Innovation Lab at Melbourne University and more recently with the Australian Food Hubs Network and the Open Food Network. Her involvement in this field dates back to my own, in 2007, though, ironically, hers began during a period of policy work within the Victorian government, while mine began at the grassroots community level in Bellingen.

I say 'ironically' because, as Kirsten recounts in the following pages, hers has been a journey out of government and, in many ways, a loss of faith in the interest of government to make the changes at the scale and urgency that are required. I have sympathy for that view and agree that the evidence for governments' willingness to achieve major food system change is (at the federal and state levels in Australia, with some notable exceptions like VicHealth) fairly thin on the ground. Yet at the same time neither of us can ignore the fact that government has the power and the resources that, should they be devoted to the appropriate programs and projects, would rapidly bring us towards the shift we all want. Where we disagree is on the likelihood of this actually happening: Kirsten thinks it highly unlikely; I stubbornly persist in believing it not unlikely, given the right combination of circumstances and political mobilisations.

This and dozens of other philosophical, political and practical issues have been the subjects of many a late-night discussion with Kirsten and her fabulous partner, Serenity. Individually each is a brilliant thinker, writer, strategist and passionate activist. Together they make a formidable and dynamic team, responsible for one of the outstanding innovations in the food systems and Fair Food movement in Australia and globally: the Open Food Network. Guided by the values and principles of open source software – that intellectual labour and creations generate innovations and improvement in code and programming – Kirsten and Serenity have formed an international collaboration of programmers and designers to facilitate the efficient, transparent and fair distribution of local produce between farmers and eaters.

And they are just getting started – so watch this space!

***

> *Our duty, as men and women, is to proceed as if limits to our ability did not exist. We are collaborators in creation.*

Pierre Teilhard de Chardin

INTRODUCTION

There are things that happen, call them whatever you want – fate, grace, synchronicity – but you know what I mean. I suspect all our lives are shaped by these 'chance' encounters. Books that jump off the shelf into your hand, email invitations that you almost deleted without reading, people and events that just happen at the right time and shape the next leg of the journey. If we listen to them, if we go where they lead, exciting things can happen.

I suspect that the most important ingredient in a meaningful life is knowing – and therefore behaving – as if what you do really matters. Somehow I have always believed that my life has meaning, and that I can live a powerful, active and beautiful life. More than that, I believe that my choices matter – not only to me, but to those around me and rippling outwards into how the world is. And more than that again – I feel that the timing and the fortune of my existence compels me to care, that my ridiculous amount of freedom and opportunity comes with responsibility somehow. This is an incredible time, so much is at stake and so much is possible.

But don't get me wrong – I know I'm just normal. I know that I think more than many, and I feel injustice and pain in the world intensely. I've always been like that as far as I know. I remember being angry when my parents' friends were racist, and I was really young and don't know how I knew I wasn't okay with it. I remember crying when I found out about slaughtered whales. I suspect I was probably quite an annoyingly over-sensitive kid (maybe still am!). But I also think that probably many children feel these things, and then learn not to.

We moved around a bit while I was a kid. I was born in New Zealand but we went back and forth to the United Kingdom a couple of times and ended up in Australia. I prided myself on flexibility and adaptability, but later realised that this affected both the way that I form relationships and my sense of connection to place. The reality of dislocation and the experience of walking into the unknown may have prepared me to understand that everything really can change overnight – and when it does, you want your people close by. It also makes for an interesting relationship with Australia, with country, which is now by both circumstance and choice firmly my home. I have fallen in love with Australia and feel my roots growing deeper and deeper with every year, but the cool, tall forests with ferns and trickling clear streams and lakes will always be 'ground' for me.

As this teenager turned into her 20s, my succession of boyfriends turned into girlfriends. This was great for me, but my parents were really not happy. It's only relevant here because it meant I had to really take responsibility for my own life. Until this point we were a pretty strong family unit, but my coming out stirred things up. I risked my family (and lost them for a while) because of who I am and who I knew I had to be. Standing up for what I believed in became intensely personal, with prejudice and fear of difference coming firmly home to roost. Principles, passions, excitement and actions all mixed up. I had to choose whether I was going to live my own life and accept responsibility for the outcomes of that choice. I chose yes, weathered the storm and it has all worked out okay. I learnt to back myself, listen to my heart and speak my truth.

'SAVE THE WORLD' PHASE 1: THE GOVERNMENT YEARS

When I was about 25, I was living in London and life was a party, a long holiday. Then one day I picked up a little book called something like *One Blue Planet* on the tube. The floodgate opened. Of course I knew that 'the environment' was in trouble, but as I sat on the tube it washed back through me like a connection to something I had always deeply known. I had no idea what I was going to do but it was time to go home and find out.

Back in Australia there was a period of searching – I had no environmental qualifications, knowledge or networks, so I read and talked to people. One of my first discoveries was this thing called the government – I'd never taken any notice of it before. I had no idea about all the things it did, tentacles everywhere, and lots of it sounded really interesting. I did go and talk to a couple of people about whether my computer engineering degree could be put to any use in making a difference but nothing resonated. So I decided that policy was a better bet.

I took a database development job at the Environment Protection Authority and accepted a government graduate position to start the following year. The position I was offered was building a website in disability services, but it felt like the right step – I figured I could wiggle my way into the Department of Sustainability and Environment (DSE) from there, which is exactly what happened. My placement in DSE coincided with the establishment for the first time of a 'sustainability unit'. I was incredibly lucky to be there in the early days of some smart and committed people grappling with what this meant. Looking back now it's quite unbelievable the doors that opened and the opportunities I was given to learn – for example, I once had a task compiling and summarising all the local government sustainability and environment policies I could find. When I was offered a permanent job at DSE I jumped at it like a piranha! 'Hurrah,' I thought, 'now I am here with the people who will work out how to save the world!'

My next job was in the newly established Sustainable Futures Unit (what the?!). No one knew what that meant ... so one day we decided to write a business plan. I googled something ('futures', perhaps) and found the Australian Foresight Institute. I read something about the possibilities of integrating Western scientific and technical capability with Eastern philosophy to chart a new pathway forward for a wiser humanity. I was moved and excited; it registered deeply. It took a couple more prods before I embarked on a Master's in Strategic Foresight. I didn't really know why, it just seemed right – and it was.

One of Ken Wilber's books, *A Theory of Everything,* notes that we sit for the first time in a position where it is possible for an individual, and a society, to access the sum total of all human knowledge. But what we struggle with are ways to integrate it, make sense of it, use it wisely to guide our decisions, understand enough about other perspectives to actually make use of it. Wilber's integral theory had a huge impact on me, teaching me how to look for and frame the different dimensions of what is going on in any issue – the physical and technical systems; the cultural and language aspects; the behaviours and practices; and the internal individual processes, morals, psychology, spirituality, etc.

Foresight aims to help groups of people or organisations to think about different timescales, break out of established patterns and expectations, wrestle with uncertainty, see patterns in systems and imagine different futures. I loved it; a new world was opening up and my capacity to think was being expanded. It was giving me a new frame through which to look at lots of things, while the things that I was struggling and working with could be seen in a different light. I became aware of a longer sense of time – thinking hundreds or thousands of years into the past or future is quite an exercise, and a useful one at a time where the short-term outlook is somewhat bleak.

This time and space gave me licence to think deeply about what was going on or what might be possible. While employed on interesting sustainability projects at work, my understanding of the problems and solutions continued to be stretched – my essays explored everything from indigenous worldviews to virtual reality technologies as psychological tools. At work, I practised clear and short writing for decision-makers. At uni, I practised saying things that were not said, or were not 'practical', 'useful' or 'possible'. I explored another 'voice' in a quiet safe space. I practised integrating things from completely different fields, bouncing them off each other, visualising and describing different possible ways forward.

Looking back I see that those years were a small window of opportunity, a chink of time when the idea of 'sustainability', and how we might attain it, was being introduced and explored. I was fortunate enough to be involved in projects where I was allowed to think across whole systems, to grapple with complex problems that no one knew what to do with, and to make space in my brain for the possibility of change. I was encouraged to think about systems change, and studying foresight at the same time gave me some tools and mental capabilities to do so. Between 2004 and 2007 it really felt like we were going to be able to do something. There was a change in the energy in those years, from an urgent but theoretical concern about climate to the realisation that the systems were destabilising fast.

So, knowing that our current systems actually aren't sustainable, then we have an increased focus on how people, communities and ecosystems can actually cope, re-form and reshape new pathways that might be, in the face of disruptive and rapid change. Somewhere in there, it started to become all about food. By 2007, there were three big wheels in motion for me:

- In the Victorian Eco-Innovation Lab (VEIL) we envisioned a design/research think-tank that could analyse, visualise and guide government towards levers of real systemic change. VEIL had a work program to conduct research reports on areas of policy interest, to analyse the issue and highlight social and technical innovations in response. We requested the first of these to be on sustainability and food.
- I was working to develop a sustainable consumption and production policy – carving out some room to move inside government on systemic sustainability policy. To explain systems and why it mattered, we wanted complex and multi-faceted examples that people could understand, relate to and care about. What better system to work with than food?

- Around this time, someone forwarded a call to me from the health department. They had been working on food and nutrition policy with the industry departments for a few years, but were increasingly realising that there was (or should be) an environmental component – would there be anyone in DSE interested? 'Sure is,' I said. 'What do you need me to do?'

But the wind was changing ... the window was closing. The political champions for sustainability moved on, and the culture within the department became ever more toxic. I started to notice how much work didn't really seem to go anywhere, like hamster wheels. Reports were left on shelves, briefings reworded until they were meaningless. We had carved out some space to work on sustainable food, but I was almost on my own. I was okay at 'the game' and progressing nicely up salary brackets, but, one day, someone who was working for me suggested an idea and my response was something like, 'Nah, that'll never work, it's been tried a million times.' Warning bells rang loudly of an emerging jaded, self-defeating public servant – time out. I was outta there.

Sustainability and systems

'Sustainability' means different things to different people. Most broadly, it means that we can keep doing what we're doing – 'it's sustainable'. In relation to the environment, it is often defined as meeting the needs of the present while not undermining abilities of future generations to meet their needs. But 'environmental sustainability' can mean everything to do with reducing impact on the environment, reducing waste, saving energy, planting trees, etc. (so, basically reducing the unsustainability of the system).

But 'systems' are more than the sum of their parts. To understand a system, you need to explore the relationships between things – not just the things themselves. So, for example, you have to consider the energy used in reducing water use, and people's responses to technology, and how financial markets intersect with energy availability.

Some argue that we're so far gone, attempting to 'sustain' our systems is actually misguided. In relation to human and natural systems, it's less about whether we can keep going the way we are (clearly not), and more about whether we can transition to systems that are actually sustainable.

Nowadays, the word sustainability often goes hand in hand with 'resilient' – resilient communities, economies, food systems, etc. Resilience is about how well a system can deal with disturbance and disruption. It is firstly about how well the system can absorb the disturbance and continue operating, or how quickly it can return to its operating state. It is also about response and reorganisation. A system can be considered resilient if it can adapt, re-form, in a new way that is 'sustainable' under the new operating conditions.

On a personal front, I'd had a good friend for a few years, a co-conspirator in DSE, and the only one who seemed to get this food thing. She was an unusual woman – I looked up to her a lot as she appeared to be on the fast track to senior executive in the public service, but was somehow so good at the game that she could choose not to play it. Serenity – in between manipulating middle-aged Labor Party men, planting ideas of Permaculture and grassroots radical revolution – had also decided that she was in love with me. Lucky for me. She'd made her intentions clear much earlier than this, but it was about now that I had a nasty break-up, a little bit of breathing time and then realised she was onto something and opened the window. If either of us was a force to be reckoned with alone, undoubtedly together we're mayhem.

The next chapters of this story are unquestionably ours – not just mine – as the decisions, the journey, the work and life itself become more and more deeply intertwined. There is no doubt that this is the story of two – our energies sometimes combining, often balancing and refining. Serenity has a way in the world that is different from mine. Her rural upbringing has her embedded in the challenges and beauty of farming in Australia. She's firmly grounded in the physical world and has no hesitation in ripping up concrete or planting a big tree. But she's so smart and committed, warm and caring and (perhaps only known to those close to her) extremely funny. Team Kerenity was born.

IT'S ALL ABOUT FOOD

I moved to VEIL in September 2007 and bunkered down reading all things food, and writing a report. By the time I came up for air I was completely hooked. The first part – the litany of problems – was overwhelming but also affirming of what I'd started to suspect, that so many of our environmental and sustainability problems come together in food, from land degradation, biodiversity loss and water use to energy and greenhouse gas emissions – and that's just in production. I started to realise that food sustainability wasn't just nice to have, that food security and sustainability were interlinked and that the natural conclusion of the current system was pretty nasty. Then the second part – the innovations – started to light me up. From regenerative agriculture, organics and Permaculture to urban agriculture, dietary changes (meat or no meat), waste, local and seasonal. Everything we need is out there and people are hard at work turning this thing around. The movement (long before I was calling it a 'movement') was stirring; the world was alive with a food revolution.

My intention with that report was to tell the story of food, to put all the information in one place and make a compelling case for change. I hadn't really thought that much about what would happen next. Lots of people supported me in this, mostly people across government who wanted this case put clearly to try and support the new agendas fighting industrial agriculture and fast food for airspace. Some of these people had started saying things about media, presenting far and wide, but I was not interested. I just wanted to write the report, give it to the government and hope that someone would do something useful with it.

Then came one of those 'things' I mentioned earlier. I got an email about a 'despair and empowerment' workshop based on the teaching of Joanna Macy. It was the next day and I just felt like I wanted to go, so I went. If you haven't ever been part of 'the work that reconnects', I recommend it to you. I'm not going to detail thc process, but basically this was the first time ever that I have been given 'permission' to really grieve for the destruction and loss that drives me – to feel that pain as physical and visceral and absolutely real. We cried and talked about anger, fear, sadness and indifference. From my public service training, 'getting emotional' does you no good at all. Put your argument together, gather your evidence and persuade someone powerful. For god's sake, don't go getting all teary. But that dark hurt place, which feels the pain of what's happening, is also where the power

– and courage – comes from. I started to learn not to be afraid of the darkness, the fear, the anger and the sadness. This was the first time I dared to look it in the face, say hello, and ask, 'What next?'

The answer came loud and clear – I had to stand up and speak for my work and the message underlying it. I had to put my face, my name, my voice and my self to it. I was introduced to the legend of the Shambhala Warrior, which resonated incredibly strongly with me but was also very personally challenging.

So I stepped away from the facelessness and non-accountability that four layers of bureaucracy gives you. I moved from trying to get someone else to say something to saying it myself. We released the report in March 2008, just as oil prices were skyrocketing and the world food crisis started to hit. I was ready to speak to it – and so the fun began!

So did that report make a difference? I think yes. The ripples travel through the system in ways you can never anticipate. Putting a message out with my name on it opened a door to a whole new world. Far beyond my intended government audience, it introduced me to people working quietly behind the scenes on farms, in projects and communities across Victoria, Australia, the world. Today we're calling them the Fair Food movement – they're the people in this book, but there are thousands and thousands of them, all over the world. People have told me that reading that report (and other things since) has changed them and what they're doing. So yes, writing reports can make a difference.

But even as I was summoning my voice and my courage and speaking out, I was still trying to get the government to act. (So then, like Gargamel in *The Smurfs,* the arch-enemy reappears.) When I was invited to come back into the public service, working in the Department of Premier and Cabinet's Climate Change Unit to try and drive a whole-of-government food policy, I said, 'No, hell no – I'm not shutting up again.' But then I read the Shambhala Warrior legend a couple more times, took a deep breath and bought a new suit. When it came to the crunch, I was still focused on policy change and had to give it another go.

I have a bit here from a journal entry that I wrote not too long into my DPC time, which captures the essence of how I feel about working 'inside'. This was written in the build-up to the Carbon Pollution Reduction Scheme (CPRS), Rudd's attempted carbon trading system, when state governments were madly making sure they could remove any other useful policy because the magical carbon price was going to fix everything:

> It's absolutely doing my head in. It is so detached from what seems real to me. On arrival I was trying to respect the perspective, seeking to learn it, and not sure what my contribution was intended (by who?) to be. But the problem is that

this perspective is completely disempowering. It is one that believes that the world is a market and can be designed and controlled as such. Even as the doctrine of free markets falls apart around us, we're busy trying to fix climate change by imagining we can design a market for this one specific part of the problem, and then correct all the market failures around it, and then correcting the market failures around those ad infinitum until we're all burnt up. Or run out of oil, or all our rivers die, or whatever happens first. It is a worldview that is attempting to design a solution that perpetuates the system that created the problem. And when you're in this worldview there's very little you can do. People have no power, only markets. People who want to do anything, create new systems or reduce their greenhouse footprint, just distort the markets and make it cheaper for brown coal generators. And government these days has no role in this – no leadership here. In the face of this certainty and complete disempowerment by both the system and the disempowerment of the people within it, I got angry and then I crumbled into depression. The government is just not going to fix this, we're quite doomed.

But I knew that anyway, didn't I, that's why I left last time, so why am I here? I have been asking myself this, because I need to know or I will go nuts. It feels like I'm meant to be here,

> and I can't believe I got the chance for no reason at all. I have been firmly ticked off by my employer: 'I did not hire you to agree with them, Kirsten – for god's sake take them on.' In discussion with my wise and wonderful partner, I'm inclined to agree that the reason we're there may be so that we can confidently tell everyone else, 'I work for the government and they definitely do not have this under control – do not wait for them to fix this.'

While working at DPC, I had also been leading food systems research projects at VEIL, ranging from Victorian food supply scenarios to an analysis of Victoria's fruit and vegetable distribution energy use and emissions. I'd continued talking, thinking, writing, learning and listening in the world – and trying to take this into policy. One of my biggest insights in this period was the power of your environment and peers to actually filter your sense of possibilities. I learnt that I could not write for VEIL when I was physically in the DPC building – I would immediately start toning it down, pulling it back, trying to make it more 'reasonable', 'realistic', 'achievable'. So I strengthened my chameleon – saying things quite differently with my different 'hats' on – making sure that neither voice got to edit the other!

Perhaps it was worth sticking it out at that point, because after a couple of frustrating years trying to drive whole-of-government food policy (that some people/departments/agendas really did *not* want to happen), there was a sudden opportunity to develop major climate-change policy in the lead-up to the 2010 state election. At moments like that it's important to have the right people 'inside' and ready to run headlong into the space that's opened up. A lot can happen in a short time when the political forces align. But, as we've seen over and over again in this country – it can be undone even faster.

TO LEARN TO SWIM, YOU HAVE TO LET GO OF THE EDGE

In 2009, the Copenhagen Climate Conference effectively failed. Then the federal government caved on climate. And in November 2010, the Victorian Brumby Labor government was toppled by a Baillieu-led Coalition. It seemed like a good time to get out of climate policy. Kirsten working inside the government was over, this time for real. There had to be another way to do this.

The road took quite a sharp turn. From a public servant to an open source software start-up. By the beginning of 2011, some things were apparent to us:

- Business as usual is not possible – our global financial, energy and governance systems are at breaking point, and it's not clear if or for how long the societies we expect can be 'sustained'. We cannot assume that surplus resources will be around to support the changes we need.

- Government policy and action are not going to reach the levels we need, *and* they are fickle and incredibly easily undone. If we want change, we have to change the structures and systems that most levels of government are heavily invested in keeping the same (local government sometimes being the exception to this).
- The world is full of incredible people who get things done. They are reshaping their communities, economies, farms and cities. They are often doing this through forms of 'social enterprise' or 'social business' – reshaping the economy with services, products and systems that can make a difference.

I wanted to get my hands dirty – I wanted to make something, do something, not try and tell someone else what to do. I wanted to be directly part of creating the world I want. Our governments don't lead, they follow – it's up to us to show the way. If they support and come along with us that's great (and very helpful), but if they don't we push ahead anyway. Serenity had another year of her PhD to do, but I was ready now – action stations. We had talked a lot about where and how things in the food movement could go next. We were connected with both a wide range of farmers and a growing community of active and interested inner-city people who wanted better, more resilient food.

We also had strong personal motivations. We wanted to do something that could support itself, reducing our personal reliance on the safety net of government jobs and research funds. We wanted to build a livelihood that let us put all our energy into activities that actively created the world we wanted to live in. Also, we wanted to be able to invest money that we earned directly into the things we thought needed financing, to bypass a lot of 'asking permission'. And we were keen to spend more time up on Serenity's family farm, learning our way around, with a view to moving there eventually.

In food systems, it seemed to us that distribution was the bottle-neck. If we want farmers to be able to make a living in sustainable or regenerative farming, and we want everyone to be able to access (and afford) the food from these farms, then distribution is where it's at. With a supermarket duopoly and increasing concentration through the food processing sector, farmers are being left with fewer options for how and where they sell their food and how much they get paid. This leaves the system brittle and far too reliant on a very small number of big players in the middle. Also, we were well aware of farmers who had food that was going to waste, often organically produced (sometimes certified, sometimes not). There is plenty of food around either from growers who are too small to warrant going to a farmers' market, or too large for it to be worth their while selling directly. Australia had quite a few 'seasonal box schemes' operating, but we knew lots of people who wanted to be able to choose from what was available and weren't interested in seasonal boxes. So we started cooking up some ideas.

We were lucky enough to be accepted into a conference in the United States in March 2011, from where we visited a bunch of really inspiring people. One of the most influential was the Oklahoma Food Co-op. We were blown away by how awesome they were, and inspired to try and adapt it into something back at home.

Then came another fateful email – this one for Gathering 2011, an event on the coming weekend in Melbourne, which was going to 'bring together leading thinkers, change-makers and collaborators from across Australia and around the world to explore what's possible, and develop ideas on how we can best solve today's most pressing social and environmental challenges, together'. On a whim I bought two tickets. Saturday morning rolled around, and Serenity and I casually rolled ourselves into the heart of Melbourne's nascent social innovation and enterprise community. Our heads exploded and we came 'home' at the same time – suddenly surrounded and embraced by a bunch of amazing people who've kept us going since. They're motivated, empowered, collaborative, getting on with things themselves – we'd never been immersed in a community actively looking for how to help each other do awesome things before. David Hood, Glenn Todd, Hailey Cooper-Rider, Theo Kitchener, Inge van der

Poel and the Hub Melbourne gang were to become our friends, advisors and co-conspirators! Michel Bauwens, from the P2P Foundation, gave us a language to talk about what we were trying to do and (later) a global community of peers who 'got it' and were on side. Sometimes magic happens.

A Fair Food system needs farmers to be able to connect directly with the potential buyers of their products, and set their own prices. People need to be able to see those farmers, who they are, where and how the food was produced. But if we don't want farmers to have to drive around and deliver all the food to each customer, or eaters to have to drive around to all the farms, we are going to need 'middle people' – they play an important role in aggregation and distribution. The problem occurs when those middle people have too much control in the system, or there aren't enough of them. We want lots and lots of diverse, locally accountable middle people helping deliver good food from farmers to other people. Let's call them food hubs.

I pondered while wandering a bit, as I do, and it seemed like we were surrounded by examples of networks that move nutrients without any central 'agent' or brain in charge. Mycelium can move nutrients through vast networks in forest floors. Slime moulds operate as individual spores until nutrients are scarce, then they join forces to make a different kind of body that can move them to somewhere with more food. The internet can move information all over the place through connected nodes that all understand the rules. Hmm. We need to build a mycelium-like network for food for us, with lots of autonomous agents making the best decisions for their local conditions, and being able to easily pass information, connect with others and join forces to effectively move food. Sure, no problem, let's do that.

What is a food hub?

Food hubs make it easier to buy and sell local food. They:

- work directly with farmers to assist in the marketing and distribution of their products, aiming to 'scale up' the distribution and market access of produce

- are built on values that set them apart from conventional systems, with many focusing on social change more than economic profit

- usually have a central focus on improving the viability of local and regional producers, with many also having a strong focus on improved sustainability and equity of access.

Given the nascence of this movement in Australia, we include local food aggregation initiatives of all shapes and sizes in our network, provided they give attention to community food security and community enterprise development.

An initial survey and desktop review has suggested that there are over 80 organisations broadly meeting this definition in the eastern states of Australia. There are probably many more. The use of the term 'food hub' is also increasing in Australia, and has been a central component of research and prototyping new models such as the South East Food Hub.

Find out more at the websites www.foodhubs.org.au and www.southeastfoodhub.org

We needed to develop and test a model that was replicable, that could be done using the spare infrastructure that's everywhere: garages, corners of warehouses, community and school halls, and potentially trips that are already being made. We wanted to simplify this process so that it was much more accessible for people to have a go at helping connect farmers and eaters. As we looked deeper into it, we found that there are lots and lots of co-ops, box schemes, buying groups, etc. And we found that many of them are struggling with the admin – it's time-consuming, complex and exhausting. But they're committed and they're sticking with it. We figured technology was a critical part of all this; surely we could develop a system like eBay for farmers ('eat-bay' was one of our first name ideas), which enabled farmers to easily communicate and sell what they have, but also enabled them to connect with diverse distribution enterprises that help aggregate and facilitate sales to local customers.

We talked to lots of those food distribution enterprises, like CERES Fair Food and Food Connect, about what we wanted to do. We discovered that software was a nightmare, even for the bigger successful ones, and that they were all sinking money (which they didn't really have) into it. Even where there were attempts to share or collaborate, it wasn't really their core business, and no one could afford to actually invest in making it shareable.

So to business. We had a grand plan, but started super small and simple. First, we started with a van, some amazing fruit and an enthusiastic network of people in the inner north. I set up a basic Zen Cart online shop and away we went. Soon we had 12 different collection locations and a lot of lessons learnt, including that the technology required for food hubs is much more than a basic e-commerce site.

Second, we started searching for software we could use – surely there must be something out there that we could pick up, use and build upon ... what we were looking for needed to be:

- transparent: enabling the customers to see the farmers, how they produced the food, how much they were getting paid, etc.
- networked: enabling farmers to sell their food through a variety of outlets, hubs and shops – so

their available inventory was drawn down through sales across the network

- diverse: we didn't want just one 'business model' for the hubs – lots of people are doing things differently, experimenting and adapting to local context. Some people do seasonal boxes, some let people choose their own, some sell in bulk, some only accept cash, etc. We felt that local identity and autonomy of the hubs was critical
- open source: this was non-negotiable. Without being able to make changes and adapt any existing system, we would always be dependent on the whims and business model of its provider. We need to experiment and change, learn from on-ground activities and develop something that worked for us – and so does everyone else
- in the Commons: ownership and motivation matter. We set up the Open Food Foundation to make this clear.

We looked, we talked to lots of people and we tried out a bunch of systems, but what we wanted simply didn't exist. It was a crazy time. We didn't think that we knew what we were doing, and as the shape of what we were going to try and do started to form it was quite terrifying.

I went through the School for Social Entrepreneurs and it was fundamental. Along with the thinking, the experimenting and the talking to people, I was also having quite a significant rewiring (some might call it a breakdown of sorts). In order to go further with any of this, I had to learn not to be an expert and that it's okay to actually not know what you're doing and whether it will work. This is the *opposite* to our dominant culture, which believes that expertise, confidence and action are about knowing the answer and how to implement it.

If we were going to do what we wanted to do, we needed to have a profound unlearning – we needed to know that we didn't know and do it anyway. We had to learn to be confident in pursuing this – because it needed to exist, we could see how it could work – even though we were extremely unclear whether we were the right people, with the right skills, at the right time. We resolved to have an ongoing test – if someone else came along who was doing this better, we'd stop and move on. We still do this test.

And it's happening. With vision, willpower, synchronicity, partnerships and incredible contributions from people here and all over the world, the grand, audacious plan of the Open Food Network is coming to life.

THE STORY BEHIND THE STORY

So now let's just go a little deeper, because not telling the story behind isn't telling the story. We all have public faces and private faces. Our public faces are generally stronger, more confident, more sure and positive than our private ones. But it's not all plain sailing, and, for me at least, there is some stuff behind being a 'leader' (in so much as being asked to contribute to this book) that I want to share.

When you act, or speak, you become exposed. All the confusions and contradictions, the questioning and the changing occurring within are synthesised into a moment of time – a decision, an action, a statement, a behaviour. From it people will impute who you are and what you stand for and, as soon as it is out there, time and your self move on. You may not feel that action truly represents you anymore, and you may feel it was wrong. But if you also feel things deeply, and wish to see change in the world you must act – and act now. I've had to learn that what I say and what I do (and today – what I write) will be okay. My job is to make sure it's authentic, that I'm really talking and acting from the best truth I know now, and deal with the cringe factor of reading it back later if I've changed my mind! So here's something I've not written publicly before.

Sometimes when I wake up, the weight of the world is too heavy. My overwhelmed mind begins the marathon through all the things I'm behind on, all the ways I'm dropping the ball, all the things I'm not doing well enough or the people I'm letting down. At the same time, this is swirled around with the things that make me feel powerless – the coal, the climate, poverty and war, asylum seekers, Tony Abbott – and the foreboding feeling that it's all going to keep getting worse – that we ain't seen nothing yet and a collapsing civilisation has much in store. And then I flick back to all the things I'm not doing well enough. If I stay in bed, this spiral can continue deeper and deeper into paralysis. If I make the fatal error of going to Facebook or the newspaper, it's quite likely I'll get lost in hopelessness and lose a few hours or maybe a day, despondently hunting for something to make me feel better. If I can catch it, and step away from it, I can usually turn it around.

I know that many people experience something like this. The 4am waves of panic and despair. The powerlessness that is so disempowering and overwhelming. Sometimes it gets all of us and sometimes it's best to distract ourselves and turn away. But it is also the cord and the connection – if I can still feel it all, and I still care about it all, then I love it all enough to keep fighting. If I didn't love, and dream, and believe so deeply that we can do better than this, I could perhaps turn away and pretend it wasn't happening. But I feel it and so there's no turning away. This is our time. This is the world we're in, the history we're part of.

This vulnerability is my strength, my courage, my determination. And something always happens. The synchronicity. A book, an email, a person – a contact from some amazing like-minded soul on the other side of the world that reconnects me with the wonder and the possibility. Vulnerability mixes with the magic, the intellect, the people and the spirit, and we get stuff done.

For me, this life is a spiritual thing. Growing up Catholic, I had experiences that have embedded very deeply in me the belief that there is more to it than what meets the eye. I remember explaining to my mum when I was about 15 that I felt closer to 'God' when I was outside and being in love with the world, walking the dog, watching a bird, listening to a tree. So instead of going to mass, I was going outside. Obviously the falling-out with organised religion continued apace from there, but I've never lost the sense that 'God' is in the world – my god is Gaia. There are forces far bigger than me at play and I can connect to them with wonder and humility. It's best done in forests, preferably barefoot.

When overwhelmed, I remind myself that I am not in charge here, I'm just a minuscule part of something way, way bigger that I can never expect to know or understand completely. One night when I was stewing about things I had to deal with the next day – how would it go, was I prepared enough – a sudden calm came over me and words rang in my head: 'This is not about you, get out of the way.' That's become a strong touch-point for me, when I'm not sure. I can stop worrying, commit myself ... and then watch what happens and adapt. I can operate from love and not fear. 'Spiritual' does not mean passive. It does not mean only contemplation – but contemplation to notice and to listen, to connect and to act.

These are things that are critical for me, but I am only just beginning to understand and explore them. The balance of commitment and letting go, of passion and detachment, of courage, humility and audacity. This journey continues to unfold.

MY GUIDING 'TIPS'

Here are some things that guide me:

- Do what you're called to do. Keep open to if and when it changes – don't get stuck thinking it's who you are.
- Dream big dreams, believe in plans, but take small steps – then watch, learn, adapt and keep moving. Know why you're acting and don't get diverted from your values.
- The power of pull – do what you need to do, and people will come with you if it's right. You don't need to try and persuade anyone.
- Don't be afraid to share your ideas and your knowledge. You're not the first person to think of it. Be open and you'll find the others – we're better together.
- Listen to the quiet voices and the uncomfortable feelings in the stomach – don't be afraid to haul your beliefs about yourself, your actions and your limitations up into the light to check if they're actually true – or useful.

- Vulnerability is power. It might make your voice shake sometimes, but it's real and that energy can connect with people.
- Take control of your finances – you can often buy yourself some freedom by learning how to get by with less money.
- Draw courage and passion from the earth, all-knowing over all our time, where wisdom comes in quietly and trees recharge the heart.

JOIN US

Humanity on this planet has been touch and go as long as we've been here. I read once that *homo sapiens* probably got down to a couple of hundred people clustered in north Africa between ice ages 100,000 years ago. And now, in our time, for the first time as humans, we're facing this choice at a planetary level – and the signs aren't great. Will we listen, learn and stand with the indigenous peoples fighting to protect the earth? Will we be able to find a way to live in balance with the planet? Perhaps. Maybe, being all connected the way we are now, we can become a planetary brain and nervous system that brings something new to life. Technology plays a part in this. And adaptability is the hallmark of humanity – the possibilities for evolution in the rapid and frequent change of the next couple of thousand years are intriguing. It might be time for another great leap.

We don't actually have more than a vague idea what we're all trying to achieve, what the world will look like in 200, 500 years from now. But it is in the search for what we believe in, how we do things, how we organise ourselves, the relationships we build, our experiments with economics and payments, that we're playing the long game ... this is how we lay the traces of cultures that might live on.

Writing this has been hard, and humiliating, in a good way. It makes me feel humble, and exposed – who am I to write about my choices like this, to 'inform and inspire'? Our grandparents went to war and gave their lives for 'freedom'. Most of us are still afraid of breaking the law or going to jail.

The Fair Food movement is one way we can choose, every single day, which world we're going to be in. And it holds millions of places for those who want to be part of the joyful, fierce work of creating it. The Open Food movement is one way of choosing – every day – to *not* be afraid of sharing our ideas, our knowledge, our tools and our lives with others. We're all in this together and, if we're going to turn this thing around, we're going to need lots more of us, and we're going to need to fight harder. But the good news is that all of this is full of the beautiful people and places that make life worth living.

I don't know what happens next, but you should come. Come alive. See you there.

A VIEW FROM 2030 – ONE POSSIBLE FUTURE

I like to describe what happened as the 'Enlivenment' – a joyful, creative unleashing of the power of passionate, connected people who discovered the power was in our own hands.

In the early years of this century, things were unwinding fast. The global financial system was creaking under the weight of accumulated debt that could never be paid in a declining economy, as were the Australian people in their mortgages, credit cards and businesses. Rude shocks just kept appearing, and fading into normality and tensions and fear rose. More and more power kept moving into the hands of the bankers, and the treadmill just kept rolling faster. Farmers were leaving the land at a record rate, taking their generations of intimate local knowledge with them. The miners, the frackers and the bankers were having a field day making more and more money off declining and dying ecosystems. The outer suburbs continued to roll, even though no one could really pay for them. The 'developers' had the governments in their hands and they seemed unstoppable. Gains from years of advocacy, policy and campaigning seemed to have failed. Pressures of everyday were mounting. Many people were scared, angry and lost.

But in and around Melbourne something special was also happening.

This strange urban Petri dish was breeding a culture that celebrated and nourished fearless, exuberant, radical experimentation and resistance. Melbourne and the regions around were full of people who were fully alive, and in love with the world they were dreaming into existence. An identity bloomed around 'world-hacking' – playful and creative disruption and redesign. This identity spread, through the suburbs, mirrored and fuelled through ever stronger connections with the country. People started to identify with it – this hive of creation and positive resistance – and with identity came courage to step outside the box and join the wave.

Once this spark was lit, resistance bred creativity. With our backs against the wall and choices narrowing, Australia unleashed the most incredible responses to what seemed like overwhelming odds.

Of course I'm biased, because this was where I poured myself – but I reckon it was the farmers and the fooders that really triggered it. We 'jumped' first – weaving connections between farmers and eaters, farmers and the land, eaters and the land, listening and learning with Australia's first peoples about how we might go forward. There is no stronger tie to a living planet than through food, and no people more strongly tied than those producing it, and we all can. All across Victoria and Australia, a wave of regenerative and transformative agriculture changed the story about what was possible, and what was good, in this country. As this story caught fire in the cities, they came alive with food. The exciting new connections between farmers and eaters dissolved the urban–rural divide. Food was the vanguard of self-determining, self-organising revolution.

The thing about food was sixfold: it reconnected people with country and a journey of working with this country in the search for real, meaningful ways to live. It required no permission from government or the corporations that controlled the food supply chain, and the ability of soil to produce food could not be suppressed. The knowledge was easily spread and the seeds of a culture of openness were already in place. Everyone could easily get involved and there was room for every skill-set – you could leave any job and be useful in this movement (even bureaucrats can be re-homed, with a little work). And it recharged and enlivened us in the darkest times – seeds, soil, deliciousness, hands, empowerment. We knew – with our feet and our hands, as well as our minds – that by working in and with this country, the future we wanted was within our reach. The sixth thing about food was that it dissolved social divides and shifted the norm around the broad need for, and the honour in, physical resistance. It was for food, freedom and country that farmers, rural people and urban people united to say both yes, we will do it like this, and no, you cannot do that here.

The second sweep of change was 'open'. Again, perhaps just my perspective, but I reckon this hit us here first through food. Seed freedom and software freedom – it seems obvious now in 2030, but until 2020 and the global shift to default GNU and creative commons licensing, people were quite comfortable with the idea that you could 'own' an idea! And it was somehow ethically defensible to do so for profit, even when that idea or technology was needed and could be improved by others. As the fight against seed patents and loss of biodiversity heated up, so did the fight against IP and locking up the ideas and technologies that we were ready and raring to use.

Melbourne was the seed of the Open Food revolution, mass development and application of technologies to reorganise and re-democratise our production and consumption systems. The embracing of technology and particularly IT to change the way we could self-organise critical supply systems – initially in food – was a breakthrough. Using these tools to enable self-organising, community-driven restructuring unleashed incredible creativity. The commitment to 'open' meant it spread and evolved, fast. We took these tools to our hearts, we trained ourselves and recruited corporate refugees and put them to work. By 2020, most of us could code, at least a little bit, and multitudes of skilled developers were liberated from pointless soul-destroying work for companies they didn't believe in.

From food the hackers moved on to guerrilla solar and wind – what a boon those smart meters were on every house. So much computing capacity able to be unleashed into community-controlled P2P power grids. In the early parts of the century people thought the Chinese were good at breaking patents and hacking the technology they wanted, but as you now know it was Australia's hacker spaces that took to solar, wind, manufacturing, processing and textiles – everything.

As well as creative hacking, we also got the hang of resistance hacking. Anonymous gave way to Synonymous as the hacker movement aligned with the need to stop major destruction. Hackers swarm well, and their ability to shut down destructive companies and systems became legendary. The error screen became every coal, gas and mining CEO's nightmare: 'Error: society has removed your licence to operate. Computer says no. Computer makes this decision Synonymously. For further information, see conscience.' Attempted development in the Galilee Basin was the first to fall ... it wasn't just about shareholders anymore, the risk was they'd take your whole company down.

Hacker swarming – so much fun. There's a real power with bodies on the line. The Occupy, Arab Spring and European resistance movements were the tip of the iceberg. With new generations of mesh networks, encryption and rapid social mobilisation, combined with the fearless and preparedness (cultivated in Melbourne's rich training grounds), we swarmed our way through the darkest years. Strategic, targeted, direct resistance – we learnt not to spread ourselves too thinly and to really disrupt. To move faster and smarter and to cover our tracks. To apply every tool we had – including technology – to slowing the destruction. It moved from a fringe activity to a social responsibility – are you in or are you out? Talk or action? Slacktivist, hacktivist, activist. Yes, people went to jail, many people.

These actions and swarms were critical to carving new and real relationships. At key sites of resistance, the non-negotiable 'over our dead bodies' commitment of indigenous peoples got inside our skin, as well as opening our eyes to what it means to resist for and with our 'country'. The stories of Jandamarra-style running resistance through country that becomes home are becoming legendary – it's already unclear what's actually 'true'. Song cycles retracing and re-weaving through Prices Point, Galilee, Maules Creek, Tarkine and Kakadu – and countless other places where we learnt to appear, to disappear and generally cause havoc. Clear calls to key sites at key times overwhelmed the corporate-government complex. Technology played its part in our newfound swarms, but many who were there swear that other forces were at play too...

Of course, all this required a big operating system reboot. It couldn't have happened without a camaraderie of fearlessness, underpinned by love and passion, fuelled by the realisation that the 'powers that be' were not going to protect these places or do these things, and we had to do it. And we could.

As more and more people sank under more and more debt, freedom from the illusion that ever-increasing wealth was either possible or good called strongly. People who did still have jobs questioned why they would spend their lives working for things they didn't believe in, and started to believe they didn't have to. In Melbourne, there were so many possibilities and people to guide into new fulfilling livelihoods that it was hard to resist. And with exciting work to do, and reasonable livelihoods being created through social enterprise and community capital, it was intoxicating – corporate refugees fled in droves.

The repression grew stronger, more and more in jail. And with it our courage and clarity – it got easier and easier to see what we were fighting. With loosened ties to 'jobs' and 'careers' and 'reputation' – fear of arrest and court cases loses its sting. The creative disruption and the resistive eruption worked together and empowered each other.

Things really got underway once we worked out how to hack money. Realising that we had enormous wealth and capital within our own communities, we consciously and interdependently reinvested in the essential infrastructure and services and court cases we needed. The superfunds never knew what hit them. Poof, we took it back. Coal deals collapsed and we found, in our communities, that we had more than enough.

So I guess really what happened was that we discovered ourselves. We stopped asking nicely, and waiting, and cajoling. We stopped trying to persuade people, beating ourselves with guilt and fear, and redirected our energy from endless paralysing evidence bases of despair. We discovered self-determination, connection, self-organisation, country and swarming.

We connected with those who cared – here and around the world – and we flexed our muscles in every sense of the word. Like adolescents testing the boundaries, sometimes we went too far, but we learnt – sometimes painfully – that there is no 'they'; 'we' are the ones who decide our future.

We opened technology up and worked together to rewire the world. We resourced ourselves and re-created money. We connected to ourselves, our land, our farmers, our food, the first people of this country, and to each other. When we were ready, we re-created governments to suit the new times.

The shadows of this possibility were evident in 2015 – as it seemed like all was lost, from Australia resounded a 'no'. A bold 'hell no'. Ideas and identity are infectious and attractive, and creative, courageous people flooded into Melbourne as a stronghold of resistance. A breeding ground, a seeding ground, a feeding ground. It became a respite, to recharge and reconnect and re-inspire, for the generation of warriors whose work and lives we celebrate today.

The world was, is and always will be, a grand tussle of ideas and possibilities – each stretching for light and trying to grab a foothold in history and culture. Twenty years ago, any significant and lasting reversal of ecological decline seemed nigh impossible, and an unstable, catastrophically out-of-control climate inevitable. The chances of this mobilisation, this revolution, this re-creation seemed remote if not impossible ... but it depends where you looked. Those who could see it, who planted the seeds of it, and who committed their minds, hands and hearts to it, they made it so.

A MOTHER ON A MISSION

FRAN MURRELL

I often describe my involvement in this movement as a cascading series of meetings and encounters. Being in the right place at the right time. Taking certain actions, and one thing leading to another. When we begin to see ourselves not as isolated individuals, but as essential parts of the whole, that is when we discover that we do indeed have wings, and we begin to learn how to use them.

So it was when my path crossed with that of Fran Murrell, co-founder of MADGE, in 2010. This was shortly after AFSA published its open letter to Tony Burke and was calling for signatories. Fran contacted me to find out who I was and where I was coming from, doing her due diligence, just as she and her colleagues have done with the biotech industry. We saw in each other a fellow radical and idealist.

Fran and her many colleagues at MADGE have done an outstanding service to this country by challenging received orthodoxy about how the new sciences of genetic engineering, nanotechnology, irradiation and others are going to 'feed the world'. Hard questions need to be asked about these matters, not least of which is: who benefits from this headlong rush to the embrace of new technologies? We should all be thankful that we have Fair Food pioneers like Fran and her colleagues at MADGE who have the courage to stand up and ask those questions.

Fran was a great support to me in the first couple of years of AFSA, serving on its interim management committee from 2012 to 2013, and attending a number of strategic retreats that led to important advances. Most notably, at Batemans Bay in February 2013, Fran joined me and five others for two memorable days where we all had the epiphany that the role of AFSA was not only to deconstruct the 'control myths' of the Big Food system, but also to generate our own 'creator myth' of the new Fair Food system that is emerging in Australia and nationally. Thus we realised that we have the power to use words and metaphor and narrative, to generate concepts and language, to create and tell our own stories. That is where, collectively, we started to test out our new wings.

***

Do your little bit of good where you are; it's those little bits of good put together that overwhelm the world.

Desmond Tutu

INTRODUCTION

I was born in London and came to Australia, for one year, after my Australian boyfriend's UK visa ran out. That was 26 years, 25 years of marriage and two children ago. We both worked in the building industry, he as an architect and me as a quantity surveyor, which was then an unusual job for a woman. We decided to start a family, and I was pretty certain I'd be back at work after six months as the baby books noted that was often when the settled baby slept through the night. Strangely, things did not quite turn out that way.

I've been described by those closest to me as opinionated, obstinate and obsessed. In my current line of activity these are very useful traits to have. A sense of humour and an ability to turn off and just enjoy the moment are also important. It is often impossible to know if what activists do is effective, useful or counterproductive, as the world frequently appears oblivious to our efforts.

Given this apparent futility, why continue? I believe that these seemingly invisible acts by people all over the world do count. Our entire society is built on the efforts of the unsung and unremembered; we inherit what they created.

Mothering, that tedious, exhilarating, mysterious and entirely ordinary role, made me realise I'm not an unencumbered and separate adult but part of society.

Like it or not, what happens within my home and my life is underpinned and enabled by what happens in my street, city, state, country and the entire world. My family flourishes best when everyone has the chance to flourish as well. I, like all of us, can make a difference. What we do creates ripples that influence others. I think we may all be surprised if we ever got to know what we have seeded, intentionally or not.

Getting involved with the food movement has challenged me. I've had to overcome barriers and expectations, many of which are in my own mind, and develop new skills. I've discovered that courage comes with practice and that 'success consists of going from failure to failure without loss of enthusiasm', as Winston Churchill once said. Perhaps the only real failure is to not see yourself as an essential part of the whole.

The people you will meet in this chapter started MADGE with the sole intention of persuading the Victorian government to extend the moratorium on growing genetically engineered (GE) canola, a type of plant breeding also known as genetic modification (GM). GE is the latest flowering of the industrial, global agriculture and food system that has developed most rapidly since the Second World War. Its proponents tell us a persuasive story: we have never had so much food, so many choices, so cheaply. The ever-increasing use of chemicals, the consequent depletion of the soil, the squeezing out of local farmers and the rise of both imports and a focus on

Australian exports is just the inevitable outcome of this perfect system, they say. According to this story, the dominance of two major supermarkets and the takeover of our food culture by processed food and fast food outlets is what 'consumers' want. To go against any of this is a Luddite rejection of progress and a refusal to bow to market discipline.

There is another, older, story that makes more sense to me. Food comes from the earth; it creates the culture of the farmers and the eaters, which in turn shapes how land is used and what is grown. As a child my father was evacuated from London to escape the bombs in the Second World War. He spent years in rural Devon and trained as a market gardener. He saw the introduction of a pesticide to kill the red spider mite. The first year nearly all the mites died, but after three years the insect problem was as bad as previously. He also read that protective clothing was supposed to be worn when spraying and realised that no one did this. This led to his disenchantment with chemical agriculture.

At that time the Devon food economy used many local resources. Farmers cleared the beaches of seaweed to put on their fields as fertiliser. The pigs were fed food scraps. They lived and were slaughtered near the main street of the local town. However, the fruit and vegetable markets were locked into distribution networks via London, which meant it was hard for my father to receive a reliable or predictable income. He moved back to London and became an accountant but never lost his love of plants. Our large garden was filled with an enormous variety of vegetables, fruit and flowers as well as chickens and a huge compost heap. I know that a well-managed garden can produce an abundance of food, that a carrot freshly pulled from the earth is divine and strawberries are intensely luscious when picked and eaten straight from the plant.

AN ESSAY TOPIC CHANGED MY LIFE

I was a young at-home mother in a foreign country, Australia, with my first baby. I had given up my full-time professional job to look after my daughter, and it was a huge culture shock. My husband worked long hours setting up his architectural practice. While I adored my daughter, I needed some kind of activity that wasn't about the daily, minute and constant needs of a baby and the vast amount of housework they create. I decided to study part-time at Melbourne University for a Graduate Diploma in Environmental Studies. It was a chance to understand more about the challenges of the world my child would grow up into. It was also a welcome return to the company of adults and it meant brief, predictable periods when I could follow my own interests, uninterrupted by the demands of my baby.

An essay topic changed my life. We had a choice to write on either forests or pesticides. The forestry debate was the one most students were attracted to and so I decided that I would work on pesticides. At this time I was a pretty conventional person. I knew that pesticides were dangerous but there were safe levels of use and they were strictly regulated. The ones that were harmful to people and the environment, like DDT, had been banned.

Imagine my surprise when my research uncovered an entirely different reality. I was shocked to find that levels of pesticides in drinking water in many parts of rural Australia would not pass World Health Organization standards. How could this happen? I found that many chemicals are released with no testing whatsoever. The tests that are done to prove 'safety' are mainly done by the chemical industry themselves. Those chemicals and their breakdown products can mix together to form even more toxic compounds. Chemicals do not sit where they are sprayed and then disappear. They circulate in the soil, water, air and food. They affect the plants, worms, soil life, bees, butterflies, birds, animals and us.

I learnt my adopted city of Melbourne had cancer and illness clusters that started appearing in the 1950s. They occurred as new housing developments spread into areas that were market gardens, and children would run and play in the sprayed fields. The clusters were also linked to the pesticides used to control weeds and termites. I discovered unborn babies and children are most vulnerable. The chemicals that mothers and fathers are exposed to can have a major effect on their children's health. Tiny quantities of pesticides or chemicals at crucial developmental times can cause harm whereas much greater quantities of the same substance at other times will have a negligible effect.

Some pesticides can be flushed out of the body in urine via the kidneys. Other types become stored in fat. Toxins accumulate up the food chain. The top of the food chain is the human baby. In the womb a foetus grows bathed in the chemicals circulating in the mother's body. When that mother breastfeeds her baby she is unwittingly feeding it the toxins that have become concentrated in her milk.

I felt really ill. I'd breastfed my darling for seven months. What had my milk contained? I had no idea what I'd been exposed to or how it might affect her. I'd also bottle-fed her until she was a year old with formula. Did this contain pesticides or other chemicals? How was it made and what was in it?

My mothers' group showed me that we don't always have much choice about how we feed our babies. One mum in my group had Crohn's disease. She'd had to feed herself via a tube down her nose during pregnancy, as she couldn't physically eat enough to keep the baby growing without this. There was no way that she could breastfeed her much loved child and so he was on bottles from day one. Another mum had twins. She managed to breastfeed them for a while but used formula too. Other mothers returned to work or chose to use formula for a variety of reasons. Some fully breastfed their kids until they were weaned. It is up to mothers to decide how to feed their children, but I felt whatever our choice is, it should be healthy for our babies. We currently have to live with not knowing whether our breastmilk contains unknown toxins or exactly how the ingredients in formula have been grown and processed.

I finished my essay on an upbeat note. I'd found out about GE crops and that they would reduce the amount of pesticides sprayed on crops. 'Hooray!' I thought. Less poisoning of babies, farmers, animals, food, land and water.

I was really curious to find out more about these GE crops and how they worked. Over the next few years I read whatever I could about them. The more I learnt the more questioning I became.

FROM GE ENTHUSIAST TO GE CRITIC

My new path formed slowly in the gaps that appeared, and then disappeared, in the daily life of raising two daughters, being a wife and running a home. It gained direction after reading an article in *The Age* about GE crops. Bob Phelps from Gene Ethics was quoted and the article gave his phone number. I rang him up and found that his office was only a few minutes' walk from my home. He had been a critic of GE crops since 1988 and had been following their development for years. I was impressed with his knowledge, open mind and integrity.

I started going round when my kids were in crèche to help with mundane things like stuffing envelopes. Bob began to ask me to do some research. He had been told about a report written by US scientist Charles Benbrook. I managed to find it online and was immensely pleased with myself. (The internet was fairly new in the late 1990s and I was amazed I could find an obscure agricultural report from the other side of the world.) This report showed that GE soy had a 5 per cent reduction in yield when compared with non-GE soy. I was puzzled. One of the claims made for the introduction of GE crops was that they would be more productive and so could 'feed the world'. How could they do that if they actually produced less?

Genetic engineering (GE) or genetic modification (GM)

GE is a type of plant breeding developed in the 1970s. Scientists take DNA from bacteria, viruses, plants and/or animals and create synthetic and chimeric genes that have never existed before to make a 'gene cassette'. This is inserted into a plant cell via two main ways:

- bacteria is used to 'infect' the plant cell
- a 'gene gun' shoots the cassette in.

The aim is to change the way a GE plant functions. The main ways a GE plant has been altered are:

• to survive being sprayed with weed killer (herbicide tolerance or HT) and/or

• to create a toxin that kills certain insects by destroying their guts (insect resistance or IR). This toxin cannot be washed off as it is within the plant.

The use of these poisons has led to the development of super-pests and superweeds, which are fought by stronger and more frequent use of pesticides and toxic seed coatings.

Newer types of GE breeding have been developed that edit genes or stop them being expressed. All these types of breeding carry the risks that:

• the new protein intended to be expressed by the GE breeding will be toxic

• GE will alter the plant genome unintentionally, causing new toxins or anti-nutrients

• the chemical sprayed on the GE plants will harm the eater.

Developments in understanding of genetics that make GE obsolete:

• The Human Genome Project showed that the idea a gene makes only one protein is wrong. They can frequently make several by themselves and also work with other genes to make one protein.

• Epigenetics shows that how our DNA is expressed depends on what we do, eat, and are exposed to.

> Genes are not Lego blocks that you can move around or alter without consequences. They are part of a 'fluid genome' that responds to what is going on around them.

This was one in a series of blows to my enthusiasm for GE crops. I'd found out that most GE crops are designed to be sprayed with weed killers like Roundup. The idea is that the weeds die and the crop survives. The other main GE trait allows the crops to produce a toxin to kill certain insects that eat it. The plant makes the toxin within its cells and so it cannot be washed off. Far from reducing pesticide use, many people claimed GE crops would increase it. Pests and weeds would evolve to repeated exposure to the weed killer and toxins in the GE crops.

Hearing these things was challenging and depressing. It punctured my illusion of salvation by technology. It went against the commonly accepted story of science as a pure discipline dedicated to improving our lives. I did not let go of this belief without a fight. I went to talks given by pro-GE speakers hoping they could allay my doubts about the technology. I heard about the talks via Gene Ethics, and I liked Bob's view that we should hear as much as we could about the subject.

Unfortunately, the pro-GE scientists, like Dr Nancy Millis and CS Prakash, were less than convincing. Their argument appeared to be that it was highly unlikely that GE crops would be harmful. However, I was not there to hear opinion but to see what research had been done, and they could not produce detailed and exhaustive studies on GE crops. I felt cheated.

I also went to talks by scientists like Dr Judy Carman and Dr Mae-Wan Ho. They showed that scientific understanding of how genes work was repeatedly undergoing revolutionary changes. That the GE crops in the fields and on our plates were developed using a now outdated view of genetics. They also showed how inadequately GE foods had been tested and that the companies that owned the GE technology had done most of the research. It was a relief to hear logical, coherent, scientific discussion of GE, but what it revealed were the yawning gaps in understanding the food that was creeping onto our plates. Could I really bear to think about what this meant?

It felt lonely and scary to be looking into a subject that most people knew nothing about. If they had heard about GE crops, they often believed they would feed the world, reduce pesticide use and create more nutritious crops. I would try to chat to people about the issue but was aware that my attempts could be counter-productive. Many people viewed GE as a scientific enterprise and therefore were only willing to consider views from scientists. I felt as if I'd stepped through the looking glass. When I looked for scientific evidence to back up the claims that GE crops were safe for us to eat, it was either non-existent or full of holes. Yet scientists in the public realm were claiming GE was the most tested food in history and anyone who thought differently was deluded. I was a mother and a non-expert and yet I felt compelled to find out. I tried to keep my mind as open as possible and look at all the evidence. However, I couldn't see this as an abstract issue. My kids were young, vulnerable, growing and dependent on me to make the best decisions for them I could. I decided I had to protect them by being cautious and yet I should find out as much as possible about this new technology in case it was beneficial.

I decided to try and produce something practical from my research. I suggested that the community crèche my children attended go GE-free as this experiment with our food would affect our children. Parents have a right to know that what they are feeding their children is different in fundamental ways from what they ate growing up. It was harder than I thought, since Australia's GE food-labelling laws are so full of loopholes that almost nothing is labelled. Some people were interested and sympathetic, but many were too busy with the daily round of children and work to have much time for debates about food. It became irrelevant when the budget cuts to crèches meant we lost the cook who had made lunch and snacks for all the children.

My children were born in the 1990s when conservative governments at state and federal levels were cutting funding to the maternal and child health service, kindergartens and community childcare. I witnessed how callously children and their carers were treated and how little they were valued. My first visit to the maternal and child health centre with my daughter was as part of a group of new mothers being told about government budget cuts to the service. Instead of being for mothers to access when they needed, it was intended to allow mothers five visits a year. Subsidies to community crèches were also being cut. As my daughters' crèche was threatened with closure, my activist skills were created and honed in the struggles to keep it afloat. The crèche did survive but only due to the unsung efforts of many people, especially the staff. The benefits of happy children, a networked community and the family support that this crèche provided dwarfed the meagre public payments that kept it going.

This struggle politicised me and made me feel that if I did not take action, then things of beauty and great value could be easily lost. In our profit-obsessed culture, children, families and their carers seemed to be regarded as a drain on economic productivity instead of the very heart of life. I understood that the real values of our society are laid bare in how we treat the voiceless and the vulnerable.

I continued volunteering with Gene Ethics when I could. Victoria introduced a five-year moratorium on growing GE canola in 2003, which was a huge win. Bob introduced me to a fellow volunteer, Glenda Lindsay. Glenda had returned to Australia after living for several years in South Africa and India. She had attended conferences on food, biodiversity and biotechnology in Africa and knew critics of GE like Vandana Shiva from India and Tewolde Berhan Gebre Egziabher from Ethiopia. She told me of the beautiful patterns that Indian women made of seeds. What looked like the creation of artwork was in fact the way the women, the traditional custodians of the seed, decided which seeds were to be collected and planted the following year. It was a combination of science, art, economics and celebration.

Many people in the food movement, in my local area and in musical circles know and love Glenda. She is a constant inspiration and a great friend. She has an enthusiasm for life and a generosity of spirit that has been incredibly important to me. She made me get out and be visible even though I would much rather sit at home and read about issues and perhaps write a submission to a regulator or an email to a politician. She is very sociable and has the ability to chat to everyone. She also is a great innovator, seeing what is needed and just going out and setting it up. From composting waste from cafes to launching a local food swap to using music to create community and healing, she has done it. Her energy and creativity have swept me into more engaged activism.

We decided to persuade our local council, Yarra, to go GE-free. Councils provide food to the elderly, childcare centres and various events both within the Council Chambers and outside. Glenda established GEFFY (GE-Free Friends of Yarra) with local friends.

We had a film fundraiser, lobbied the council and got a GE-free clause in the council's catering contract. It was great to have an event and meet others who were interested in finding out about food. It was exciting to get a tangible result from our actions with the council clause. However, we did not have the time or ability to ensure that everyone understood and enforced the decision. Perhaps this can only be the case when everyone knows what GE food is and where it is in our food.

GEFFY also used the money to buy seeds, and we did three days of food mandalas at the Sustainable Living Festival in Federation Square in 2003. The first day we used black pepperberries, red rosella flowers, yellow bunya nuts and lemon myrtle flowers to create the Aboriginal flag. We surrounded it with the rainbow serpent made up of the DNA spiral. It was a celebration of foods and plants traditionally eaten by Indigenous Australians and the intricate mystery of genetics that supports it all. The next day we created a different mandala of the wheat, beans, lentils, coffee, potatoes, apples, lemons, rosemary, quince and cinnamon that the white settlers brought. The final day we combined the Indigenous and white settler foods with chillies, pineapples, star fruit and more from Asia and the Pacific.

It took five hours each day to create the mandala, and at the end we would unmake it and use the seeds for the following day. It was a great way to talk to people about food and the importance of seeds, diversity and knowledge. We had many great conversations. While we were creating the first mandala, a workman who was repairing the spire of the cathedral adjacent to the square visited us. He had travelled all over Australia and knew the bush foods. It showed me knowledge is all around us, and that we just need the right circumstances to allow it to appear. We also spoke to a farmer. She had just got off the train at Flinders Street and spoke to us of the effect of the drought on her and others in the country. It was drizzling as we worked, and it made me realise how separated we are in the city from the people and the land that feed us. We cannot let this estrangement continue because we need each other. One child was so fascinated with what we were doing that he insisted his parents bring him every day. Other people loved the designs and were amazed at the beauty and variety of the seeds and food. Making art of the ordinary allowed us to see that it is, in fact, miraculous. Talking to others and sharing seems to me to be the way to help a considered and considerate society to evolve. Listening has taught me so much.

Working with Glenda was creative and fun. It is hard, if not impossible, to challenge the status quo without a friend or two who understand, who can be there to laugh at absurdities, commiserate over difficulties and show the next step when it seems a brick wall has been reached. One huge inspiration for me was the *Cities Feeding People* conference that Glenda helped organise. It was exciting to hear Jackie French talk about using hardy trees to create a grove of plants that provided cool shade in a harsh climate. Jude Fanton spoke about the importance of seed saving. Visits were organised to community gardens on local public-housing estates. These diverse and productive gardens flourished within the urban landscape of Collingwood and Fitzroy. It was encouraging to see people finding generous and elegant ways to solve problems by working creatively and respectfully with the natural world.

I continued researching GE, studying at university and mothering my kids. The interruptions from life and children were so great that it took me a decade to finish the Grad. Dip. I changed from Environmental Studies to Urban Planning and Design. I found the destruction of the natural world by our economy too confronting, and thought that making cities friendly to kids, food and living was more encouraging and hopeful. I aimed to return to work in a new career as a planner. I just needed my kids to grow up a bit.

I volunteered in the school canteen when my kids were at primary school. When I first started, a couple of children were acutely allergic to nuts. By the time my children left primary school, so many children were allergic to so many foods that the canteen manager shut the canteen when she went on holiday, because she was concerned that one of us might sell an allergic child a food that could kill them.

I was really shocked at what I was seeing. As a child I had one friend who was allergic to nuts and my sister's friend was a coeliac and that was it. Why were so many children ill and why were there not major inquiries into food? Why were the genetic changes to our food that could only occur in a laboratory not being discussed as a potential cause of this ill health? Why was this major increase in food allergies among children being brushed off and downplayed?

I continued researching and going to meetings on GE that were attended by farmers, food producers, activists and politicians. It was enlivening to realise that I could be part of an important discussion rather than an apathetic bystander. I worked hard at trying to understand the complexities of seed, the food system, politics, regulation, patents and markets. It seemed that one thread, GE, was part of a complex tapestry of issues.

THE BIRTH OF MADGE

On Mother's Day, 13 May 2007, *The Sunday Age* reported the Bracks government would review the moratorium on growing GE canola and either extend it or allow it to lapse in 2008. I felt instinctively that working to preserve the moratorium was vital, and I decided to take up the challenge.

Soon afterwards I came home to a message on my phone's answer machine from someone I'd never heard of. She enthusiastically announced herself as Jessica

Harrison and said that Bob Phelps had given her my number. She suggested that we dress up as cows and stand outside the United Dairy Farmers of Victoria (UDV) meeting in June. The dairy industry had been instrumental in the original GE moratorium being passed, as they had not wanted to feed their cows GE canola. Her phone message made me laugh out loud as she was so cheerful and energetic.

Glenda, Jessica and I met and decided to call ourselves MADGE – Mothers Are Demystifying Genetic Engineering – after the group Mothers Against Genetic Engineering that had been so effective in stopping GE animals and plants in New Zealand. We felt that women need to know what is in food and how it has been produced. They are the major food buyers in Australian households, do most of the cooking and frequently nurse sick children and family members.

I had no idea how to start, but Glenda and Jessica quickly whipped up a media release. It announced MADGE would be at the UDV meeting and invited others to dress up as cows and join us. Small things can mark decisive points and those few paragraphs sent out to the press gave birth to MADGE.

It felt weird that we had created something out of nothing and now we were being contacted for comment. I was relieved and impressed when Glenda eloquently conducted a phone interview with the ABC while crossing Spring Street on the way to our demonstration. The UDV meeting was in the Sofitel

Hotel in Collins Street, so we gathered in the food court of the adjacent shopping centre. Amazingly enough, several people had hired cow costumes and come along even though they had never met us. It was exhilarating to see others were serious enough about food to look silly. We stood outside in our costumes, handing out leaflets, chatting to people and hoping that someone would listen.

After the farmers had entered the meeting we went to have a coffee and debrief. When I heard the meeting had voted to allow the lifting of the moratorium, I was shocked and saddened. I knew it was optimistic to expect farmers to listen to a bunch of women in cow suits. However, it felt wrong that Victoria now risked being polluted by GE canola, against the knowledge and wishes of most of its citizens, according to polling data. I sat at the table feeling angry and decided that Monsanto and its GE canola had upset me so much that I would do what I could to stop it.

To ensure MADGE survived beyond this one outing, we decided we needed a website. I met Lindy Churches, a designer who was also a young mother. She helped me put together a site that explained what GE was and why people might be concerned about it. I listened very carefully to her feedback as I'd been immersed in the topic for years, whereas she brought a fresh and inquiring eye to the subject. When the site was finished, she said she'd like to be the first MADGE on the database, so I was very encouraged.

I decided that although the site had a lot of useful information, no one would repeatedly visit it and so we began writing a weekly digest of GE news that was sent via email to anyone who signed up.

Glenda, Jessica and I worked hard to get people to lobby their state politicians to extend the moratorium. We had a stall at the Organic Expo, produced a free postcard that was distributed in cafes and wrote a submission for the GE Crop Review Panel. We were joined by people we'd met at the cow demonstration and by others who found us along the way.

Two young women farmers were incredibly helpful. Leanne heard about our UDV demo and had dressed up in a cow suit in Horsham. The local paper printed her photo on the front page. She was concerned about the chemicals her husband used on the farm, and her research into GE had worried her even more. She invited Jessica and me to stay at her farm and organised for us to speak about our concerns about GE at a public meeting in town.

I had always been terrified of public speaking, and the first time I gave a talk, I read from a sheet of paper with a wobbly voice and shaking hands. However, the sky didn't fall, and afterwards it was interesting to chat to people and learn more about the realities of living and farming in rural Victoria. Some of the benefits of being a volunteer activist are the great hospitality we have been shown by farmers and supporters all over the country, as well as the opportunity to learn more about the people and land that feed us.

Becky, a biodynamic farmer from Gippsland and mother of three, helped research the MADGE submission to the GE review panel. Jessica, Glenda, Becky and I presented to the panel our carefully argued and substantiated report into the economic benefit of Victoria retaining the GE moratorium. The panel was only able to consider the financial aspects of the GE moratorium, although the public assumed it was an investigation into the safety of GE food because a scientist, Sir Gustav Nossal, chaired the panel.

Many people and groups, including MADGE, Gene Ethics and Greenpeace, worked together to lobby politicians to renew the moratorium. We knew that we were having success when one told us that if he received 12 phone calls, letters or emails on a subject it showed the public was concerned. He had received 600 messages from his constituents asking him to extend the moratorium. We organised rallies for the steps of Parliament House. People helped out in all sorts of ways and also spread the word in their communities.

I was hopeful that the moratorium would be extended because the work by so many people had led to a backbench revolt in the Labor Party. However, Steve Bracks resigned and the pro-GE John Brumby took over as premier. His government wanted to position

Victoria as a biotech state and appeared to believe that it needed to grow GE canola if its biotechnology health research was to be taken seriously.

THE GE MORATORIUM IS LIFTED

The Brumby government announced the lifting of the GE moratorium at the end of November 2007. It was a bitter lesson to me in how our democracy becomes thin and vulnerable under the pressure of powerful and connected vested interests. The experience was repeated in 2010 when the growing of GE canola was allowed by the Western Australian government despite similar opposition. This made me both despair and also acknowledge the importance of politics and of supporting politicians and parties who are willing to stand up for their constituents.

GE canola could not be planted until February 2008. Japanese buyers came to Australia to ask us not to grow GE canola. Several Canadian farmers, including Percy Schmeiser, who was sued by Monsanto for GE canola that had established itself on his farm, toured Victoria explaining how GE had caused contamination and market loss, and was tearing apart previously harmonious rural communities. It was great to meet people from Japan and Canada, although it was difficult to hear about the damage being caused by the pushing of GE-patented agriculture onto a resistant public. It emphasised to me that GE is about the control of food and farming by global corporations that regard us, and the rich diversity of life, as profit opportunities.

After the moratorium expired, I could have decided that I'd done enough and retired from the unpaid, constant work of activism. I did not do this because every day I had to provide meals for my family. I couldn't forget what I knew. I would have had to lie to myself and abandon my kids and their friends to a risky future. Instead, I decided to work with the others in creating a safe, transparent and rich food system as well as showing why the mirages of GE were leading us into a dead end.

The most important lesson I have learnt is that listening, co-creation and collaboration are essential. When people, who are generous enough to volunteer their time and energy, are free to follow their passion it brings amazing results. This means that MADGE has looked at many different aspects of the GE issue and uses a variety of methods to let people know what we've discovered.

Research has always been the major focus of our activities. It is important to make sure it is as accurate and relevant as possible. I have spent many hours on this but the assistance of others has been crucial.

I rang Madeleine Love, who was standing as the What Women Want candidate in the 2007 federal election, to ask her their policy on GE food. She hadn't heard of GE coming to Australia. After she was unsuccessful at the election, she became the most incredible researcher. With a science degree and professional work history in actuarial science, she read constantly, widely and deeply across the scientific literature and contacted scientists and regulators. She spent months reading the data that Monsanto had given Food Standards Australia New Zealand (FSANZ) for the approval of its GE canola, GT73, which we grow and eat in Australia.

GE crops have their DNA altered so they will create new proteins that change how the plant functions. Proteins can cause allergic reactions; therefore, the new GE proteins are compared with a database of known allergens to see if there is a similarity. Madeleine discovered that Monsanto was not able to identify and characterise one of the intended proteins in the GE canola. Instead, it supplied FSANZ with information about a protein that was not expected to be in the GE crop.

She also discovered that Monsanto had provided feed for various animal tests done by themselves or collaborating bodies. There were serious peculiarities in the way the feed was grown and prepared. In one case the GE and non-GE canola had been grown next to each other. The plants had not been 'bagged' and therefore it is highly likely the pollen of the two types of plants would have contaminated each other. This would make trials with animals fed these GE and non-GE crops irrelevant, as neither feed would be pure. The GE crops were cooked at unusually high temperatures when the animal feed was produced. There were gaps in the reports of the deaths and removal of animals from studies. The studies looked at the size of chicken breasts or the tenderness of lamb chops from animals fed GE feed. It was hard to see how these tests were relevant to mothers deciding if they wanted to buy GE food and prepare it for their families.

MOTHERS WILL NOT BE SILENCED!

Madeleine wrote reports and many press releases detailing these and other findings. I learnt that, despite Madeleine being one of the only people in Australia to have read the GT73 data, she was not listened to. She was a mother in a mothers' group and was therefore considered credible only in relation to parenting. It did not appear to matter that scientists publicly supporting the safety of GE crops specialised in insects, weeds or microbiology rather than human health. It was disappointing that epidemiologists like Dr Judy Carman were not asked by the media to comment on Madeleine's work.

We put out our findings in the MADGE digest, in press releases and reports and put them all up on the website. Lindy's sister, Renaee, took this task over from Lindy, who had originally set up the site, when she had a new baby. Renaee is creative in finding ways to improve the site and is very patient with the vast learning curve on computers that I have still not mastered.

Others have helped too. Lisa researched the protein powders her son was using and got the Victorian health department involved. Rachel wrote a concise submission to the Food Labelling Review that showed her understanding of consumer issues and research abilities. She persuaded me to go to a media training workshop that has proved very useful. Eva is a quiet source of support and inspiration and has been a committee member for years. She was one of the MADGEs who spent five hours occupying a Melbourne supermarket. Greenpeace had tested a soy baby formula and repeatedly found traces of GE contamination in it. We put signs and tape over the product on the supermarket shelves to warn buyers of the lack of labelling. Channel 7 filmed the mums from MADGE and Greenpeace in action. It takes courage to be willing to step out of ordinary life and be interviewed on national TV.

We always try to make our events friendly, informative and fun. On 4 July 2013 we set up afternoon tea on a table on the pavement outside the Australian office of Monsanto on St Kilda Road in Melbourne. We did this in support of US mums who have formed Moms Across America. Their children were sick, with allergies, autoimmune diseases and a range of other conditions. They'd found that their children's health dramatically improved once GE food was taken out of the diet. They encouraged other mums to use the Independence

Day parades in the United States to raise the issue of GE foods.

We put up our table, laid it with homemade scones and gluten-free chocolate cake and began handing out our flyers, designed as an afternoon tea menu, to passers-by. We were pleasantly surprised that others came to join us even though there was a howling wind blowing. Our fellow tea-timers included Lisa, a woman having great success using and spreading the word about regenerative agriculture both inside Australia and internationally, and Tova, a mum who has since joined our committee.

We have done several events outside Monsanto's office, and it is interesting to note that they often shut the office when we arrive. Security guards and police are frequently in evidence. It makes us feel impressively powerful that this global corporation is afraid of a few mums, grandmas and other members of the general public. This is a great lesson and I think we should note that when ordinary people ask polite questions, the corporations hide.

We have run many stalls, given talks and supermarket tours, had picnics and even printed shopping bags with the MADGE logo and 'OMG GMO WTF', which have proved very popular. We know that many people are not able to come to our events, but they help in seeking out non-GE food, questioning companies about their ingredients and setting up their own food-growing gardens.

The industrial food system has created a desire for many to find out what they are eating by getting to know their farmers and food suppliers. Katie has been excellent at showing me the work of chefs and foodies that are supporting and expanding the local food network. She is also brilliant at finding ways to communicate our ideas.

This is a taste of what people have contributed to MADGE. Some turn up to one event, others work with us for a short time and some stay for years, but all are valued. I hope that everyone enjoys the experience of volunteering with us, no matter how long they participate.

Meeting all these people has been one of the great joys of MADGE. The most important for me is Jessica. No one can beat her energy and enthusiasm. She has developed an incredible network of contacts throughout rural and urban communities, which means she is very knowledgeable about the effects of GE on the ground in Australia. She has been with me throughout the whole MADGE journey, for which I am deeply grateful.

IT'S TIME TO QUESTION AUTHORITY

We are being told that where and how food is grown and processed is irrelevant. All that matters is if it looks good, it's cheap and we like the taste. The Australian public were alerted to the dangers of this system when frozen berries grown in Chile and China and exported to Australia were found to cause hepatitis A due to faecal contamination. This happened at the height of berry season in Victoria. Similarly, onions from the United States were found to be contaminated with *C. diff,* an extremely nasty bacteria that causes severe gut infections. This was despite the local Australian onions being cheaper than the imported ones.

There is ample evidence that pesticides are harmful and that we as a society are paying the cost of these chemicals. We are eating GE food despite much evidence that it causes harm, as Steven Druker sets out in his book *Altered Genes, Twisted Truth.* GE crop breeding has expanded due to the ability of companies to control and profit from seeds based on the legal decision that genes are patentable. Patents undermine the ability of farmers and gardeners to benefit from the work of their predecessors and profoundly alter the basis of agriculture. Suddenly, corporations are allowed to profit from the countless generations of crossbreeding and selection of crops that have created the food we eat today.

In Australia 20 foods, including apples and tomatoes, will soon be irradiated to kill fruit flies. This process is the equivalent of exposing food to 2 million X-rays and it destroys nutrients as it 'cooks' the food. We will not know the food has been treated, as it looks the same as non-irradiated food. The labelling requirement for irradiation is likely to be removed soon.

Nanotech ingredients are currently in our food and food packaging but are unlabelled and untested. Pesticides and fertilisers as well as personal care products can contain nanoparticles. Synthetic biology, a field that has been called GE on steroids, has developed a synthetic vanilla. A 3D printer prints out DNA designed in a computer. This is engineered into yeast that is put into vats and fed with sugar to produce synthetic vanilla. The US Food and Drug Administration has decided that this is a natural process that therefore does not require any safety testing or labelling. Real vanilla is harvested from orchid vines growing in forests in Madagascar and Mexico and provides a good income for many small farmers. Their future has not been considered. It appears the ability to profit from experimental and risky technologies outweighs all other considerations including our health.

Is it progress to use an increasing amount of new technologies from corporate laboratories to change what we eat in invisible ways? Governments, regulators and standards bodies allow these technologies to be used. They frequently base their decisions on data provided to them by the companies that stand to benefit from selling these chemicals, products and GE seeds. The risks are borne by us. It is time for eaters to become less trusting and more questioning.

WHY MOVEMENTS MATTER

It can seem that everything is too hard and we should all just shut up and shop. However, the industrial system only exists if we refuse to see beyond its threadbare promises. Just as the little boy pointed out that the emperor wore no clothes, we need to replace the poisonous illusion of industrial agriculture with regenerative farming and food.

Currently, Western governments and global agribusiness giants are planning to 'feed Africa' by taking over the existing small farms and replacing them with industrial agriculture using GE seeds. They seem unaware of the numerous detailed, well-researched reports showing that agroecology is the only way to feed us all, reduce rural poverty and cool the climate. It relies on locally based adaption to conditions, creative and scientific problem-solving without resorting to expensive inputs. It can double yields, increase biodiversity and create the happiness and prosperity that come from a secure food supply. The global Food Sovereignty movement uses this type of agriculture.

What is agroecology?

Agroecology is a whole system approach to growing food, feed and fibre that:

• uses design to require few chemical and energy inputs, instead relying on ecological interactions to create their own soil fertility, productivity and crop protection

• enhances farmers' skills, allowing them greater freedom to solve problems

• cannot be patented or controlled by corporations.

It has already created huge benefits:

• The System of Rice Intensification (SRI), developed by a French priest in Madagascar, has doubled yields by adding compost, increasing the space between plants, using less water and turning weeds into the soil. It has been successfully adapted to other crops including wheat and maize.

• According to the UN report *Agroecology and the right to food,* 'agroecology, if sufficiently supported, can double food production in entire regions within ten years while mitigating climate change and alleviating rural poverty'.

All over Australia and the world people are working to re-establish a food system that puts human wellbeing at the centre. This grassroots movement depends on people deciding to be part of it and working to create the legal, political and economic structure to support it.

None of us can change this food system on our own. For MADGE, being a member of the GM-Free Australia Alliance is crucial. GE-Free New Zealand is a member and OGM Pacifique, based in New Caledonia, has recently joined us. We collaborate with farmers, scientists and civil society organisations all over the world. It is a beautiful thing to be part of.

Linking up with people who share the values of fairness, good food and farming as well as caring for nature and people is a source of inspiration, information and strength. We all have an opportunity to be part of this network and help it grow.

My advice to readers, based on my experience, is to find something that is meaningful to you and begin to expand its place in your life. It can be as small as growing a basil plant on your windowsill or making a soup and inviting your neighbour to share it. Try to approach life with kindness and good humour, towards yourself and others. Rest when you need to and celebrate often.

FROM COUNCIL DOG CATCHER TO CRITICAL FOOD SCHOLAR

CAROL RICHARDS

Carol and I just missed each other in the early 1990s: Carol left Manchester for Australia in 1990, while I arrived in Manchester from Australia in 1993. Sixteen years later and on the very opposite side of the world, we found ourselves sharing an adjacent room during the 2009 Agri-Food Research Network conference in Auckland. Carol was already an accomplished sociologist; I was a newbie PhD student, presenting on a thesis topic that was only just starting to take shape.

The Agri-Food Research Network brings together students, scholars and, increasingly, practitioners from many institutions across Australia and New Zealand. Carol became convenor of the Network that year, serving the Agri-Food research community for the following four years.

Fast-forward to early 2012 and Carol was one of the members of the first AFSA national committee. In March 2012, she invited me to join her on an Australia-first research project, investigating the potential of urban agriculture to meet climate change, food security and urban resilience challenges in the Australian context. In addition to providing much-needed income for a struggling PhD student, that research led to me applying for and obtaining a life-changing Churchill Fellowship.

When the People's Food Plan process launched in September 2012, Carol soon came on board to work with me to synthesise the feedback from the 40 public forums and 600 participants into a working paper, which we launched during Joel Salatin's speech at the Sydney Town Hall in February 2013. Carol also took part in the strategic retreat at Batemans Bay that framed the Fair Food movement. Carol has played a seminal role in the development of this movement, and it's been a joy and a privilege to work with her through its initial phases.

***

> *It's not the size of the dog in the fight, it's the size of the fight in the dog.*
>
> Mark Twain

GROWING UP IN THATCHER'S BRITAIN

There are so many reasons to be involved in a fight for justice in contemporary Australia: the 'stop the boats' asylum-seeker policy; gender inequality – just look at our current Australian federal cabinet with only two women; marriage equality; and the environmental consequences of coal and coal seam gas extraction. So why food? Isn't it innocuous, neutral, benign, even? With food plentiful in Australia, and readily available to most people, you might think that food justice advocates who 'make a meal' out of food are just looking for injustices anywhere and everywhere. But as an academic researcher and a member of the AFSA national committee (2012–14), the more I look at food, the more I see it as a site of resistance for many of our social ills.

While I have not always seen social, economic and environmental injustices through the lens of food, I have always seen injustices. Growing up in post-industrial Manchester in the 1970s and 1980s, as a member of the 'unwashed masses' during the Thatcher regime, a lass could not be complacent while human rights were eroded. This wearing down of rights came from all directions and conspired to keep the poor and vulnerable in their place as unions were broken and the factories closed. By the time I was old enough to leave school at 16, the earlier elation of Britain's first woman prime minister had not only diminished but turned into despair, hopelessness and working-class self-hatred.

My school had done its best with its resources and in the context of the social epoch it occupied. The 'careers teacher' told me to aim for two O levels (subject-based qualifications) to secure a spot for myself in a typing pool of the emerging service industry of banks and insurance. I ended up with one O level, in English Literature. I failed 'English Language', to the shock of my dad who said, 'You *are* bloody English!' I am not thick – I have a PhD now – but like the rest of the so-called undeserving poor at that time, there was little to aim for, little encouragement to excel and no jobs for school leavers. Our fathers, who had held down the same job for 20 years in trades and manufacturing, were finding themselves out of work. Our mothers were working – in part-time, cash-in-hand, unskilled, casual jobs such as cleaning or caring. I was one of the lucky ones, spending school holidays shadowing my mum in her various 'bits and bobs' of jobs that were known for paying 'pin money'.

I did eventually make it into the workforce on a Thatcher-style Youth Training Scheme (YTS), previously known as the Youth Opportunity Program (YOP). YOP was quickly bastardised to 'YOB', in reference to juvenile delinquent 'yobbos'. I landed at a veterinary clinic, and the government paid me 25 quid a week to help my boss and his enterprising small business 'stimulate the economy'. My peers and I essentially worked for little more than the dole, and carried the label of worthless yobs. The boss thought that the other YOBs and I were low-class scum, and treated us accordingly until we were spat out at the end and replaced by a new batch. A hunger for justice –

something shared by millions who for various reasons cannot tolerate inequality based on power, class or accidents of birth – was something I gained early in my life.

After a run of odd jobs, I landed a job as a Manchester City Council dog warden. My mother was proud. A council job was sought after and, her words, I could 'earn the same as a man'. It was probably one of the best jobs I ever had. I had my own van, a list of jobs and the freedom to plan my day. I was the only woman in a team of eight, a novelty in some respects, but which also made sexual harassment a regular feature. Still, I felt streetwise, and reasonably in control, even though I was bitten a few times, thumped, and shot at once when venturing too near a disputed drug territory in pursuit of an injured dog.

I felt like I was busting out of the well-defined gender categories. I loved the camaraderie of the union and set off to London in a mini-van to march against Thatcher's poll tax and a host of other Tory ills. But when the announcement was made that the dog warden service was to be privatised, I took it as a sign to move on. I did not want to be employed by a private security company tendering for short-term contracts, and certainly not after the Tories had abolished the minimum wage.

A NEW LIFE IN AUSTRALIA

The Thatcherite policies were oppressive, Manchester was gloomy and run-down from the lack of investment and the weather just added further insult. By 1990, I had a permanent resident visa for Australia and a one-way ticket to Melbourne. Although not as gloomy as Manchester, Australia was still recovering from the 'recession we had to have' and jobs again seemed elusive, especially for a foreigner whose skill base was catching packs of stray dogs. There was no stray dog problem in urban Australia, so I made holy bread at a convent, cared for guide dogs in a breeding and training facility and even spent a few days touting for work for a sleazy tow-truck business owner, who suggested I would do better if I wore more see-through dresses! By the time I found myself in north Queensland, selling post-mix soft drinks to tourists for a bungy-jump company, I knew I needed an education for the chance of a better job, to break the pattern of downward mobility.

By the age of 24, I had married Russell, a tall, dry-witted Kiwi. I was ready to tackle some higher education despite not doing so well at lower education – hopefully in an area that would allow me to make a difference to the world I inhabited. Not quite fathoming how to do this, I enrolled in a 'community welfare' degree program in far north Queensland. I hated it. I felt like I had joined the ranks of the middle classes – it was all butcher's paper and dispensing welfare at arm's length. My key turning point came with the arrival of Dr Roger Wilkinson, who was flown from the Townsville campus weekly to teach sociology. Bingo! This was what I was after. He taught us about McDonaldisation and the rationalisation of everything from clothing retail to factory farming. I was hooked on sociology – it spoke to me loudly.

Sociology is the study of the social world. It examines the way in which social structures – such as gender, power, governance, politics and the economy – can impact upon actions, and how people make sense of the world in everyday life. Sociology engenders a 'critical' approach, which means not always accepting the social world as a 'given' but examining inherent power relations. For instance, examining the relations of power exerted through the dominance of major corporations in the global food system. For me, sociology also helped to explain the British class system, social inequalities around race and gender (including sexual harassment in the workplace) and how power is maintained and reproduced over time. These concepts apply equally to 1980s Manchester as they do to understanding how poor people in Global South nations are relieved of their land by transnational finance corporations.

After transferring to the University of Queensland to pursue a double major in sociology, I graduated with first-class honours and was launched into the paid world of justice. Quite literally. My next gig was at the Queensland Department of Justice, researching juvenile crime and state responses – but true to form, I couldn't hack the conservative state politics, and management didn't appreciate my research, which showed that Indigenous children thought juvenile detention was a good choice for them, when it was supposed to act as a disincentive to criminal behaviour. One day, I saw that my boss had doodled axes on my report about how the current system was disadvantaging Indigenous kids and, before I knew it, I was reapplying for my own job. I didn't get it. I was idealistic, I guess, and hadn't counted on dismissal by stealth for reporting the bleeding obvious. After a few months, I applied to go back to university in the pursuit of a PhD.

LEARNING ABOUT FOOD, RESOURCES AND GLOBAL INEQUALITIES

Although not the heady days of the Whitlam era (bless Whitlam) for higher education, my university debt from my undergraduate days was manageable in the sense that I would die of old age before I paid it off, and the PhD program was free. I went back to university in 2002 and continued working part-time at a women's refuge that had been the mainstay of my undergraduate years – there's that justice theme again! My research topic, quite by accident and by virtue of an advertised scholarship, was in the sociology of sustainable grazing – that is, examining the link between society, culture, economy and natural resource-management practices. I had a ball travelling around remote Queensland areas, speaking to graziers about the beef industry, their position in the global marketplace and land management practices. This was my entrée into food and an academic career.

I learnt a few things during those PhD days:

- that broad-scale, export-oriented production systems (such as wheat and beef) are dependent on mass vegetation clearing, chemicals, capital and machinery, and are subject to the vagaries of the global marketplace (with individual farmers in this system not having much of a say in how they farm the land)
- that farmers and graziers work hard to produce food but often end up in debt and are locked in to a system that is essentially a 'race to the bottom' for the environment and farm livelihoods
- that the 'blame the victim' approach ignores the structural inequalities for debt-ridden farmers (which is similar to British YOBs being blamed for their unemployment).

Academics are like farmers in some ways. Our industry is often reduced to inputs and outputs. Chemicals in, wheat out. Grants in, papers out. Although still tenuously employed through a series of fixed-term positions in academia, I occupy a privileged position. I have ticked the metrics boxes, achieved the correct inputs and outputs and managed to stay employable.

It's not a bad thing that those of us reliant on public monies have to account for how it is spent, but like many a scholar, I have to question how useful it is publishing in peer-reviewed journals that are hidden behind online paywalls. At best, these are read by my colleagues and peers, and maybe a family member who wonders what it is I actually do. It satisfies the 'publish or perish' mantra of universities, but ultimately, is it enough to aim for?

THE ROLE OF AN ACADEMIC AS A 'PUBLIC INTELLECTUAL'

For me, it is not enough to know about the problems with the global food system and share that only with a few intellectuals. Food, its production and trade, is at the heart of many injustices: hunger, exploitation, dispossession and environmental degradation. There is another way to produce food, de-coupled from corporate vested interests, and re-democratised into diverse systems of production and ownership. But how do we help support these other systems? Not through writing for a narrow audience but by joining up with other like-minded folk: academics, farmers, food heroes, social entrepreneurs, community workers, etc. – a bottom-up rather than top-down approach. Food is vital to life, health and environment. So why would we defer responsibility to a corporate sector in pursuit of profit rather than social and environmental goods? It doesn't add up.

While universities run workshops on how to publish in peer-reviewed journals, academics are rarely taught how to disseminate research findings beyond the academy, or, importantly, how to effect change. It is often said our work can be 'picked up by policy makers' as though governance is inspired by evidence rather than politics, economics and, quite often, cronyism. Like many academics, I am winging it in this space, but learning, thanks to the dogged determination of like-minded folk who are also blessed or cursed with a similar, justice-seeking radar. Like many academics, I am trained in my discipline, but not in campaigning, politics or advocacy. It has not necessarily been encouraged in Australian universities either.

Indeed, I wear the badge of 'change agent' or, worse, 'activist' rather awkwardly. Why are the right-wing lobbyists pursuing vested interest agendas never called activists? It is often used in a pejorative sense, alongside 'agitator' and 'trouble-maker'. Isn't expecting justice something 'normal'? Is this not what we all seek? Let's remove the gaze from the so-called activist and instead place labels on those who are motivated by greed and power, actively seeking and maintaining powerful positions through corrupt 'mates' networks', opaque dealings and strategic behaviours aimed at elevating oneself through the misery of others. It is these behaviours that are behind mass environmental destruction and income inequality, and these behaviours that should have earned pejorative labels.

This is where my academic work and community work intersect. In recent years I have researched the industrialisation of the food system, economic globalisation, food production, regulation and governance. Most of my work has examined the power relationships in food systems, from global land grabs where financial investors acquire vast tracts of land from underneath 'peasant' subsistence farmers in the Global South to how supermarkets exert power along the supply chain. More on these matters later, but for now, it is useful to highlight the food system inequalities, and the reasons that anyone with access to such knowledge has a moral obligation to act.

GETTING YOUR HANDS DIRTY

About the time that I started my PhD, Russ the dashing Kiwi and I also bought a little house near the university. It was the first time I had owned a garden. I had heard of Permaculture and started to learn how to grow our own food – and that's a steep learning curve that I am still traversing. It didn't go that well in the early days; every fruit tree we planted yielded only one piece of fruit. One avocado. One nectarine. One apple. We were under siege by possums, fruit flies and eventually a fox that took our chickens and others in the neighbourhood. This was also during the extended drought period in Brisbane. An urban farmer's life is tough, and gives me an appreciation for the 'real' farmers growing food.

In parallel with growing food, I was also becoming increasingly concerned about the chemicals, animal welfare issues and various unlabelled junk that ends up in the food we buy at the supermarket. These include gassing and ripening agents, genetically modified foods, irradiation, chemical herbicides and pesticides as well as a range of 'E numbers' and other ingredients only understood by chemistry graduates. This is the end point of the industrial food system, and where we, as often hapless, uninformed consumers, buy our food in the belief that the government would not let us eat anything harmful. There is an implicit trust in our food system that has not been earned, but assumed, largely as Australia has not had a major food scare such as mad cow disease, which blighted the United Kingdom.

Avoiding buying food from Australia's supermarket duopoly is intensely satisfying, and even sporting. Those that have discovered this delightful insurrection have also realised how radical it is to find alternative sources of food, such as growing, swapping, sharing, box schemes, gleaning, dumpster diving, community-supported agriculture and other forms of buying direct from farmers via farmers' markets or the Open Food Network website.

Growing a lettuce is quite a political act in itself. In this current political climate, acting outside of the formal, corporate-controlled economy is heavily discouraged. This rattles the cage of the corporate beast and their governments (the ones that used to be ours) to the extent that some US jurisdictions have banned growing food in private gardens and chemical company Monsanto has mobilised its political and economic power and fought against the labelling of genetically modified foods. Similarly, in 2014, AusVeg, the peak body for the Australian vegetable industry, issued a press release supporting the government's withdrawal of the community gardens grant, declaring that community food growing is a biosecurity risk.

Taking control of the food system, expecting and asserting rights over the food we eat, is political. This is largely due to the conflation of corporate and government interests, and the potential for the food movement to grow and divert profit away from the corporate sector, as journalist Felicity Lawrence has recently pointed out. She argues that such is the dissatisfaction with the major supermarket chains in the United Kingdom, civil society is reverting back to more traditional forms of sourcing food, through farmers' markets, family businesses, etc., and the supermarkets have taken a major hit on their profits. We are exerting the same pressure here in Australia, though we have a way to go yet. But understanding the political landscape around food, and problematising food and eating, is a good step forward.

Why growing lettuce is an act of resistance:

- it represents a reskilling of individuals and community

- it reconnects humans and the land, reminding us of our origins and past abilities to farm and grow healthy food
- it reclaims the right of self-sufficiency – it's 'off the grid' for food rather than energy
- growing a surplus and sharing or exchanging are steps towards an alternative economy
- engaging in informal economies reconnects humans with each other and teaches us that we don't have to accept the reliance on governments or major corporations
- it's a step in the direction of greater food sovereignty
- if you grow enough lettuce that can be exchanged for other goods and services, the potential for increased autonomy over paid work patterns is raised
- lettuce is good for you, and belies value-capture through added sugar, salt and fats ... it's the opposite of the junk food industry.

Clearly, lettuce is a metaphor for any food that can be produced and distributed by civil society rather than big business. There are a number of people now who have cut down their hours in formal employment as they have entered an informal economy around food. Getting your hands dirty is horticulturally similar to your grandparents' day, the methods and techniques for growing food are the same, but the meaning has changed. They were a part of a continuum of food-growing populations, whereas we let go of that skill and quality, and our basic food has been replaced by 30,000 varieties of supermarket pre-packaged, additive-laden food-like substances. As food writer Michael Pollan has pointed out, we shouldn't eat anything our grandmother wouldn't recognise as food.

TRANSFORMING THE SYSTEM

Growing and swapping food in the community is one thing, but how do we work towards transforming a 'food system'? With global economic integration, air travel and free-trade agreements, the food we eat is increasingly the product of a global market. It is very much a food 'system', characterised by lengthy, anonymous supply chains. I recently enquired into the origins of the squid for sale at one of the few remaining fishmongers in Brisbane. It was caught in New Zealand, processed in China and shipped to Australia. This is not ideal, but possibly reflected in the price, and at least the small-scale business had an understanding of the food it sells.

If we want local food, local jobs and local supply chains, we have to pay. We have become used to cheap food, a habit that is going to be hard to break. This 'food miles' story is more straightforward than most. Like the bacon in the supermarkets made from 'local and imported ingredients'. What does that mean?

It means that those with the least power in the food system (farmers or eaters) have to accept what the lobbying power of big business permits on the label. This is why we don't see the names of the agrichemicals, gassing agents, genetically modified ingredients (with some exceptions) on food labels.

There is a lot missing from the food system: transparency, accountability, fair returns for farmers, health impacts and much more. Without this information, every eater needs to be a full-time researcher to understand the food that they are eating. Yet eating is among the most intimate of acts – we ingest food, it can make us well, and it can make us very sick too. This lack of transparency in food readily translates into a lack of awareness of food. It is convenient for the food industry for people not to know about the pink sludge that is chicken nuggets, or the ripening retardants that extend the shelf life of apples but also deplete the vitamin C content.

This is where food movements come in – shining a torch into the darker corners of the food industry, acting as a counterweight to the vast GM or junk food PR machines, and organising society towards better solutions, either through grassroots actions like reclaiming the kerbside, community composting schemes, community gardens, buyers' groups, food swaps, etc. or through lobbying governments and industries. There is a lot to be done in this space, and a lot of people who want to do it. For me, it's the perfect space to apply my knowledge of the food industry and see if collectively we can make positive changes to the world through food.

Inspired by the incredible Jeanette Longfield from an organisation called Sustain UK, I and some other folk tried to replicate the organisation in Queensland, Australia. Jeanette is dynamite – a quick-thinking, fast-talking bundle of intelligent fun who for many years grew the food movement in the United Kingdom from Sustain's London headquarters. Visiting there was like being given an endless box of fair trade, soft-centred, organic chocolates. The variety of amazing humans beavering away at their desks, transforming the food system through lobbying, research and hard work, was a sight to behold. Campaigns for 'real' bread, organic food, sustainable fish and community gardens were fought and won from that office – and when Jeanette visited Brisbane in 2009 to share her insights into the Fair Food movement, I had to act.

A group of Fair Food visionaries came together to work out how to effect change. Networks were created and ideas forged, but ultimately it did not take off. Few were experienced in movement building, we were already ultra busy and active in other areas, and in the end, it slid from our grasp and we donated our kitty of $13 to another organisation. The will was there, but we could not quite work out the way forward at that stage. I don't think this false start was a bad thing, in hindsight. We had to meet and explore the parameters of this issue, our place within it and what we could do.

A few months after Sustain Qld folded, AFSA was emerging following a meeting in Sydney. It was gathering support and momentum around dissatisfaction with the federal government's National Food Plan. The plan was an abomination. Business-as-usual was on the cards, or worse, a corporate food plan that overlooked many of the most fundamental aspects of food: ownership, health, justice, equality and diversity. A number of us contributed to a submission, and later learnt that the consultative process was quite closed with some sessions being invite-only. We took exception to the steering group, which was mainly made up of industry stakeholders without civil society representatives. After a conversation among some of the early AFSA members, we decided that we could not effect change through such tedious 'pseudo-consultation' processes with a government that was cosy with big business, and that our movement would not only have a grassroots flavour, but also aim to achieve change by investing less time in engaging with government and devoting more time to effecting change – that is, identify what is needed and 'just do it', and, in some key areas, AFSA has achieved this.

Many individuals and organisations were doing amazing things, often ad hoc, uncoordinated, and at the local level. Through the national organisation of AFSA, we hoped to get a Fair Food movement 'flying in formation' and recognised as a genuine alternative to the corporate food system. Through amazing souls like Cat Green, a Fair Food organiser in Brisbane, and legions of other folk, a series of kitchen-table talks were organised across the country, information gathered, collated and analysed, and a People's Food Plan was produced as a counter to the bland and uninspiring, and now defunct National Food Plan. At AFSA we were proud of our crowdsourced-model food policy document, filling the gaps the government had left by inserting people, humans, communities, animals, environment, and concepts such as fair trade and the preservation of agricultural land into the 'preferred policy' landscape. These are things that you would hope were just 'normal' or 'obvious', but under the current economic and political system of free-market neoliberalism, these are considered to be non-tradeable public goods. They relate to society rather than economy, and so were not factored into official approaches to developing food policy.

TIME TO REFLECT – WHAT HAS BEEN ACHIEVED?

At the height of my involvement with AFSA, I was also the convenor of the Agri-Food Research Network as mentioned above, and many of my Fair Food friends and colleagues were members of both groups. For me this provided an extraordinarily fruitful space to connect the knowledge generated from research into local and global food systems, and all the bits in between, and work with a bunch of smart and motivated people to work out how to translate this understanding into action. From my perspective, AFSA grew so quickly we barely had the capacity to keep up. The metaphor of the duck gracefully gliding across a still pond comes to mind – it looks good from afar, but there was some frantic paddling under the surface. I am really grateful to the hardworking people that I shared that journey with, and proud of some of the things we achieved with barely any resources, while also juggling demanding jobs, maintaining families and relationships and aiming to challenge a highly unsatisfactory food system.

Despite having no budget, no paid staff and no premises, AFSA has achieved quite a lot:

- the organisation of Fair Food Week with zero budget for the first time in 2013, with over 15,000 people attending 112 events nationally
- support from famous people with a social conscience who worked with us to promote the movement, including US singer Jack Johnson, who provided AFSA with a stall at the Sydney Opera House during his gig, and Australia's rugby union captain David Pocock
- the production of the People's Food Plan in response to the government's National Food Plan

- the crowdsourcing and funding of Australia's first ever Fair Food documentary
- a seat at the UN Food and Agriculture Organisation's International Civil Society Mechanism on Food Security and Nutrition thanks to AFSA president 2012–14 Michael Croft (see section entitled "THE ACCIDENTAL FOOD SOVEREIGNTY ACTIVIST")
- the formation of the Fair Food Farmers United branch of AFSA, thanks to the tenacity of farmer and current AFSA president Tammi Jonas (see section entitled "THE VEGETARIAN TURNED PIG-FARMING BUTCHER")
- the establishment of AFSA branches in Brisbane and Adelaide – with more to come online (see Cat Green's chapter about Fair Food Brisbane, see section entitled "THE RADICAL HOMEMAKER").

One of the challenges of community work is volunteer burnout. Many of us reported challenges in this regard and identified the need to step aside for a while, catch up with paid work, family duties and life in general. Luckily, the success of the movement has meant there has been a steady stream of keen and interested people to take up the challenge of maintaining the trajectory of the food movement – and there is the promise of more achievements ahead.

For me, a break from committee duties at AFSA means time to catch up on emerging issues relating to global food and agriculture. This includes continuing to explore the impacts of global land acquisition – or land grabs. Land grabs are a major issue of our times, with land increasingly viewed as an 'asset class' and therefore subject to investment by financial institutions such as banks or pension funds. Informal land tenure systems in many Global South nations mean that land is vulnerable to foreign ownership, and is often acquired under the guise of 'food security' amid claims that land is 'under-utilised'. The International Land Coalition (ILC) reports that countries whose land has been the subject of foreign ownership and control include some of the least economically developed: South Sudan and Papua New Guinea (around 4 million hectares each), Indonesia (3.5 million hectares) and the Democratic Republic of Congo, Mozambique, Liberia, Sudan and Sierra Leone (each with 1–3 million hectares under foreign ownership). While Africa is host to the largest proportion of known investments, it is closely followed by Asia and Latin America. The ILC estimates that between the years of 2000 and 2011, large-scale land deals accounted for 203 million hectares of land. The acquisition of land by financial institutions and investors has resulted in the forced eviction of peasant farmers in many parts of the world, compromising their livelihoods and making them vulnerable to food insecurity. As Oxfam has reported, Australia is also involved, with its four major banks – Westpac, ANZ, Commonwealth and NAB – funding land grabs in countries such as Papua New Guinea and Cambodia.

Unfortunately, these neo-colonial land takeovers are funded by big business, through mechanisms of the type we are told to embrace under the free-market economy. The cautionary tale here is not to trust free-market proponents, as there are clear winners and losers when the 'market' is not a level playing field in the first place. The lack of power of the dispossessed means that their experiences are not widely reported, but this is changing thanks to the work of NGOs such as the International Land Coalition, Oxfam, The Transnational Institute, GRAIN, Food First and Oakland Institute.

THE FUTURE – DIVESTMENT IS PEOPLE POWER!

My next mission in environmental sociology is also linked to finance capital. With colleagues from Australia and the United Kingdom, I am researching the fossil-fuel divestment movement, which is gaining ground globally. It's related to agriculture and food in the sense of land-use conflict; the mining of food producing land; its contribution to climate change; and again, the power imbalances when determining the preservation of the global commons. In this sense, the atmosphere, water and food security are globally shared goods and services, yet potentially compromised due to the economics of carbon extraction.

Divestment involves the socially motivated withdrawal of capital by public and private investors from corporations in protest at unethical business interests or practices, which it aims to change by exerting financial and reputational pressure on the company. To date, the divestment movement has been successful in mobilising disparate and distant populations to either withdraw their investments or pressure organisations such as churches, superannuation funds, universities, banks and cities to do the same. This is interesting in the sense that it is coordinated, widespread and has mainstreamed resistance to corporations involved in environmental destruction. The other curious aspect of divestment as a social movement is the power of lone operators to effect change – such the case of the Whitsundays tourism operator who travelled to the AGM of Deutsche Bank in Frankfurt to describe the impacts of the expansion of the Abbott Point coal terminal on the Great Barrier Reef. Similarly, by placing pressure on banks, superfunds and universities, individuals can meaningfully and legally engage in resistance to the actions of unruly corporations. The beautiful irony is that the globally integrated economy is so far reaching that anyone with a bank account, superannuation fund or university enrolment can exercise some level of economic power, and collectively, so much more – and in effect, use the tools of neoliberalism (free markets, individual responsibility, capital) against itself. It's a sociologist's fairy tale, hopefully with a happy ending.

THE VEGETARIAN TURNED PIG-FARMING BUTCHER

TAMMI JONAS

By 2013, Fair Food was becoming a recognisable movement in Australia. There had, of course, been a food movement for many years previous, but it was rather disparate, fragmented and highly localised. In February 2013, there was a 'gathering of the food tribes' at Donkey Wheel House in Bourke Street, Melbourne, at an Eco City Food Forum, funded by the City of Melbourne and facilitated by the budding 'foodpreneur' David Hood.

This is where I met Tammi Jonas. There are few other people I have met – and perhaps not coincidentally, a number of them are also from the United States – who have the energy, passion, and sheer capacity for hard work that I have observed in Tammi in the two years we've known each other. To have created a highly innovative – and, above all, viable – model of small-scale farming in barely three years, in a country where more than 70 per cent of farmers depend on off-farm income, is little short of miraculous.

But it's no miracle, of course. It's attributable to the creativity and commitment that Tammi, her husband,

Stuart, and their three children have demonstrated since establishing Jonai Farms back in 2011.

Some months after we had first met, Tammi confessed that she was an activist at heart, and she wanted the farm to be a platform for advocating a fairer food system. Knowing her track record with crowdfunding, I asked her to help crowdfund the Fair Food documentary in September and October 2013. We organised a couple of one-day retreats at Jonai Farms in February and March 2014, and from those two days we established Fair Food Farmers United – the FFFU – as a producers' chapter of AFSA. We were again exercising our power to create, and build our narrative.

Tammi joined the AFSA committee in May 2014 and stood for president in November 2014. Now she finds herself at the helm of a national food movement that is rapidly rising in profile and prominence. She also runs a successful free-range pig farm, with on-farm butchery and farm shop, charcuterie, a year-long program of workshops, interns, regular speaking and media engagements ... the list goes on. How she manages it all is beyond me, but I'm truly grateful she does.

***

> *I awake each morning torn between a desire to save the world, and to savour the world. This makes it hard to plan my day.*

EB White

A POLITICAL AND VEGETARIAN AWAKENING

I grew up on a diet of wholly industrial food. I thought I hated broccoli because it only came frozen in a plastic bag that Mum would drop into boiling water until the entire house stank like the depths of Ronald McDonald's colon. I was convinced pork chops were more closely related to cardboard than pigs (which wasn't far off given the conditions under which those pigs were grown and what Mum did to cook the fear of trichinosis out of them).

With Ruffles potato chips for entrée and Twinkies for main, Wonder Bread was pretty much health food in our house. Add a piece of Oscar Mayer bologna, some Best Foods mayo, unwrap a slice of Kraft cheese and you were looking at my most regular meal aside from Cheerios. I knew the names of everything I ate, or rather the names of the mega-corporations who manufactured my 'food'.

Much of this diet was highly processed meat products like bologna (which apparently originates from mortadella but loses a lot in the industrial translation) and hot dogs, so altered from their original state it never really crossed my mind that they came from animals, let alone how those animals were raised.

I don't blame my folks for this upbringing – their own food knowledge was pretty limited. They came of age when tinned food was hailed as a miracle in preserving maximum nutrient bang for your buck and working parents felt they at least had some options in a busy life. And while they didn't teach us good food habits, they made up for it in other ways: regular quizzing on civic-minded decision-making and clear expectations to be disciplined, self-motivated individuals.

In high school I told my coach that no matter how much I trained to run hurdles, I never seemed to develop stamina. Coach asked me what I was having for lunch. 'Powdered doughnuts and Cherry 7 Up, why?' I asked.

'Christ, Woods, no wonder you can't breathe after 100 metres!' He insisted I eat something healthier at lunchtime. So I upgraded to microwave burritos from the corner store.

But looking back I can see there were glimmers of awareness that there was food for flavour and not just fuel. I was like a blowfly at friends' houses around dinnertime in hopes of a meal made with actual ingredients instead of pulled out of a box or a can (when dinner was served at all). Somehow I worked out that pizzas with tinned mushrooms were virtually inedible, and became 'that person' who always asked for fresh mushrooms or none.

I was vaguely interested in cooking too, though it was like being vaguely interested in astrophysics, for all I knew of the topic. A major achievement in my final year of high school was for me to regularly bake eggs in bologna cups in a muffin tin as an afterschool snack, still never thinking of the thousands of chickens and pigs in the ammonia air of their sheds in the Midwest.

When I left home for the University of California, San Diego (UCSD), I actually looked forward to the sad salads and bizarre meat dishes served at our cafeteria, because at least they were regular meals.

In 1990, during my third year at UCSD, the United States was gearing up to invade Iraq in the guise of being good global citizens. I had just started to take note of politics and America's foreign policy, and got myself along to a standing-room only public lecture with a stellar line-up of academics and activists who systematically debunked the government's excuses for preparing to attack.

Step into my shoes for a minute – I am the daughter of a proud former marine (two tours in Vietnam) and LAPD cop. For 12 years of school I held my hand over my heart as I pledged allegiance to the flag, and the national anthem could bring tears to my eyes at the start of a baseball game. I loved my country just like Daddy and the news taught me to, and I vaguely recall uttering, 'Just nuke 'em,' when hearing of an alleged threat from Iran in the late 1980s.

But there in that auditorium as I stood up the back in my acid-washed jeans, leather bomber jacket and glam-rock permed hair, I could feel the foundations of my ill-formed conservative ideology crumbling around me, leaving me embarrassed at all the social markers distinguishing me for what I was – an ignorant, privileged white girl from the sticks.

I rushed home to my share house to offload 20 years of guilt on my progressive housemates, who listened sympathetically and welcomed me into the fold at last. The next morning I woke up galvanised to do something useful, something to atone for my years of complicity with the US military-industrial complex. I looked up the student union and found myself in the office of Andy, the red-afroed editor of the student newspaper, a mid-20s ratbag activist sitting at the genius end of the spectrum, destined never to finish his bachelor's degree.

Andy listened to my story and naive horror at what I'd discovered about our government and the people we'd been oppressing around the world for generations. And he helped me formulate a plan – I would fast in solidarity with the people of Iraq and ask everyone to join me. Flyers were made and posted, and to my great surprise I suddenly found myself at the centre of the student activist movement – the UCSD Alliance for Peace was born.

We chose a fast because the sanctions our government was laying on the region meant thousands or millions of people in Iraq were already feeling the pangs of hunger as food distribution was thwarted and disrupted. I wanted to feel their hunger and speak their pain since they were rendered voiceless in our oblivious world. For someone who hadn't thought a lot about food, it emerged very clearly as a critical lens to focus attention on others in need.

I only lasted until day two of the fast, when we conducted a die-in across our 1600-acre campus, and I ended up dry retching in the heat by the time we reached the Price Centre for speeches. But the protest carried on for the entire 43 days of the war and would somewhat ironically change my life for the better.

Of course, for the people of Iraq, the war didn't end after 43 days, nor did it improve their lives. Direct action quite often has no detectable impact aside from awareness raising in the broader community, and some of that can be a negative awareness when protesters 'inconvenience' others trying to just go about their day. On the other hand, the lives of protesters are quite often transformed forever, as the power of the collective is experienced and a commitment to social and environmental justice is made. And, typically, we each go on to continue to work for a fairer world – to deepen our understanding of society, nature and our place in it.

Like any nascent leftist intellectual, I started devouring information about how society is organised and my consciousness was not only awakened, it was on fire. Many of my new friends were vegetarians, so of course Peter Singer was on the reading list, and, like so many others before and after me, *Animal Liberation* left me with no real option but to forswear all meat. My political awakening found its focus in food once again, and it's never stopped, but this time my solidarity was with non-human animals.

When people would ask me why I was vegetarian, I would say it was for three reasons: ethical, environmental and health (because cholesterol was the devil in the 1990s, and margarine was the answer). More than anything, though, it was the eye-popping horror I was feeling at what I'd read about intensive livestock farming.

I'd learnt of sows confined in cages their entire lives for fear they'd squash a piglet or fight with another sow during gestation. I learnt that poultry and pigs in those sheds have chronic respiratory issues due to breathing the intense ammonia air of their enclosed sheds (and that one producer told Singer: 'at least I get to get out of here at night. The pigs don't, so we have to keep them on tetracycline, which really does help control the problem.').

I learnt that everything I'd been eating was actually the taste of fear, suffering and hopelessness. And in 1991 there was virtually no such thing as 'free range' unless you were raising your own animals somewhere, so the only option for me was to simply stop eating meat.

I succeeded in avoiding all meat until a fateful restaurant encounter after an oft-teary few months as a vegetarian back-packing across the 23 languages of Europe. There I was in Geneva with my Australian lover, who would become my husband, on our last night in Switzerland before an overnight train to Spain. Being the Californian nomad that I was, I ordered tortellini with my last Swiss francs with total confidence that it would be filled with the goodness of cheese, preferably three types. When the little parcels arrived I noted the darkness of their fillings and asked Stuart to test them for me. 'Yep, meat, probably veal,' he solemnly declared. My heart sank but my stomach shouted and my wallet was empty. I not only ate meat that night, I probably ate a baby animal raised in even more dubious conditions than its adult counterparts.

From that night I still ate mostly vegetarian but stopped using louder English to try to convince the good people of Europe that chicken stock was not in fact vegetarian. By the time I returned to the United States, I'd slipped back into eating conveniently, though continued to avoid most pork and poultry.

FINDING MY PLACE AS AN ACTIVIST OMNIVORE

With my heart now beating towards Australia, I fast-tracked the remaining two years of my bachelor's degree and hopped on a plane clutching a fiancée visa in 1993, marrying Stuart by the end of that year. I entered a family who pride themselves on a wholefoods diet, which brought about a whole new level of understanding food and the role of three regular meals per day, both nutritionally and socially.

However, I became profoundly depoliticised in my early years in Oz as I melded my identity from a tie-dyed hippy activist California girl to Stuart's conservative, leafy suburban upbringing, desperate to fit in to a family full of traditions. My carefully constructed Country Road persona found acceptance much more readily than my bolshier real self (a word I learnt here – in America I was just like every other candid feminist intellectual, but in Oz I was a 'bolshy American female'). I ate meat, I enjoyed the spoils of privilege, and the furthest I pushed my politics was the odd dinner debate with Stuart's politically conservative family.

Activism re-emerged as a *raison d'être* during my education degree in 1996, when I organised my cohort to write a lengthy letter of complaint about the poor standards we were finding in the course. How dare they expect so little of those who would teach Australia's young? We workshopped the letter with the support of a sympathetic professor who was teaching us the importance of participatory democracy – that is, real, lived democracy where you participate in determining the kind of world you live in – and stunned the university with the group action. We received a letter from the department, a lecturer took marks off my work for submitting a feminist reading of an assigned pop psychology text rather than summarising its alleged wisdoms about why stepmothers nag, but a few things changed for the better before I graduated.

I was offered a job teaching English at one of Melbourne's most prestigious girls' schools the same year I returned to vegetarianism. My politics were creeping back and quietly subverting my conservative dress code. I do cringe a little when I recall dogmatically declaring to some of my new colleagues that I was ideologically opposed to zoos as they tried to tell me of their excursion plans...

During my three years as a high school teacher I had the incredible opportunity to work as the deputy director of a remote campus owned by the school, which ran an environmental education curriculum. I found myself immersed in the complexities of logging and destruction of old-growth forests for woodchips, and living on a property that was also a working farm. Girls were taught everything from how to help a struggling cow give birth (it involves sticking your arm somewhere urban 14-year-olds have never considered) to the glorious, hardworking independence of carrying all your needs on your back for a four-day expedition while carefully exploring fragile ecosystems.

I questioned my continued resistance to eating meat in the face of the cattle raised on those paddocks and killed for the farm residents' consumption, just as some of the girls faced their food for the first time and flipped to join me with a vegetarian diet. But the chicken still came in regular shipments from Coles, and the questioning didn't result in a return to omnivorism. On the other hand, that year certainly whet my appetite to be back on the land, though it would be more than a decade before we would realise that aspiration.

A few years later we found ourselves in Santa Cruz, California, a haven for progressive politics and organic food. For the first time in a decade I rediscovered my people, and they were granola-crunching, Birkenstock-wearing lovers of blue skies and clean water. I shed the last vestiges of my poorly fitting tepid social conservatism, donned purple corduroys and suede Birks, and wore my first-born close to my heart. And it was there I truly rekindled a deep engagement with wanting a better world for our and everyone's children.

I made it through my second pregnancy still a healthy vegetarian, high on the goodness of fresh, organic greens from Santa Cruz's two weekly farmers' markets. Everyone from firefighters to stay-at-home mums, rubbish truck drivers to surfers talked endlessly about the provenance of their food and our fortune to be in an organic nirvana.

While breastfeeding my second-born, Antigone, I learnt of an acquaintance's shocking experience of being asked to leave a lolly shop for breastfeeding her baby in the store. Outraged, I quickly organised a feed-in out front of the shop in protest. Some 40 or 50 mums turned up and fed our babies on the footpath in solidarity with the wronged mother and baby, peacefully demonstrating that a baby's right to feed is far more important than a shop-owner's offence taken at a woman's breast.

Next I found myself feeding Antigone on community television as I expounded further on every woman's right to feed her baby in public, and every baby's right to feed in public. Food Sovereignty wasn't a phrase I was familiar with yet, but I see that action as the first protest I was explicitly involved in for it.

Another two years found us back in Melbourne and pregnant with our third child. Unlike my healthy experience during the first two pregnancies, I was laid low with severe anaemia by the end of the first trimester, and no matter how many supplements I took, I couldn't shake the chronic lethargy. My alabaster skin was a clear sign of oxygen-deprived blood that I tried assiduously to ignore.

Sitting at work one day, staring listlessly at the screen, I had a sudden thought that 'a burger would fix this'. Telling nobody, I sidled down Smith Street, Fitzroy, to the local takeaway and ordered a burger, hold the sauce, and can you please add some spicy English mustard? After seven years of no meat, many asked me whether it tasted bad or made me feel ill, but in truth, it was the most glorious thing I'd eaten in a long time, and I felt better almost immediately.

Red meat and fish became two to three times a week staples again, much to the benefit of my colouring and capacity to walk 100 metres without losing my breath. I didn't look back – I knew this was a forever reversion to omnivorism. And yet still I worried about the provenance of the meat I bought for my family or ordered from menus as varied as Subway and Ezard.

Returning to meat, what I did notice, which I'd had no reason to for some years, was how scant the information given to consumers about where their meat comes from really is. As a vegetarian, I didn't have to worry about it, but now that I was regularly purchasing meat, I was troubled by a lack of clear labelling, or the vague assurances from waiters that 'um, yeah, I'm pretty sure that's free range'.

Pork and poultry were still off the list for me for a further two years until I found Lilydale chicken. The initial excitement of finding free-range chicken that was widely available was quickly dampened when I investigated a bit further and discovered the many variations on the meaning of 'free range'. Stocking densities, de-beaking, sub-therapeutic antibiotics ... how in the world is a person to eat their ethics when the system is so obfuscated?

Around 2006 I commenced a PhD at the University of Melbourne. I chose to look at the role people's engagements with the ingredients, implements and cultures around 'other' cultural foods plays in the development of a more cosmopolitan society. As an urban dweller, my focus was predictably on practices of consumption, as I was quite far removed from production, aside from our series of rather farm-like backyards.

At the same time, I started a food blog – *Tammi Tasting Terroir* (now *Tammi Jonas: Food Ethics*) – initially as a repository in the cloud for my musings about food and culture as I worked across many different home, work and study computers. Before long, the blog became a serious point of connection for me with others concerned about food for any number of reasons – family, culture, cultural capital – but I'll return to the blog and its impact on my connectedness, understanding and movement building later.

Postgraduate life enveloped me and shaped me as a researcher and critical thinker with a deep understanding of 'foodways': a set of social, economic and cultural practices around the production and consumption of food. As I immersed myself in cosmopolitan theory, I began to untangle why food is central to everything – it is sustenance, it is culture, and, in terms especially of global movements of food, it is power (or lack thereof).

My cultural lenses became increasingly political as an examination of household practices around food exposed the structures of power that so often pre-determine how we will encounter <insert 'ethnic'

food> for the first time. For example, if you're of the generation where Vietnam is just a brutal, sad war seasoned with two decades of new migrants who transformed the streets of Richmond and Footscray and brought us *Romper Stomper* and a fear of dark train stations late at night, the *pho* might not taste as good as it will for the generation who have grown up in a world of gastronomic multiculturalism divorced of its specific and limiting genealogies.

As an understanding of the political nature of food was growing through my research, so were my explicitly political activities, and so I was elected campus president for the Graduate Student Association (GSA) at Melbourne Uni. During my time with the GSA, I organised a soup kitchen protest on behalf of the poorly supported tutors, mostly PhD candidates on small stipends (if they were lucky), who were doing countless numbers of unpaid hours in their efforts to educate the next generation of undergrads (we don't need to mention university executives' salaries, right?). Even in higher education, food was the vessel to make a political point about poverty and power.

In 2010, I was elected national president of the Council of Australian Postgraduate Associations; my desire to fight for any system's underclass certainly wasn't waning. A year spent traversing the country to meet with postgrads, politicians and university administrators put me in the position of making a lot of compromised food choices at one cafe after another. As I honed my negotiation skills, so did I let my food standards drop. When you're hungry at the airport before the last flight out of Sydney, there aren't a lot of choices.

And yet when the egg and bacon roll beckoned, I couldn't forget the taste of a million pigs' misery, and I ate the soggy salad sandwich instead. In truth, I ate a lot of nuts that year (easy protein). But also chips, and a pretty awful meat pie on the train from Newcastle to Sydney.

Before I even finished my stint in postgrad student politics, I knew I was fighting for the future of a sector in which I no longer desired to work. Managerialism, research conducted on the basis of insecure three-year grant cycles, an increasingly blurry focus on serving the global public good, and too much time spent in the company of vice-chancellors were enough to kill my interest in a career in higher education.

Those years of student politics, however, helped me find a way to put my activism and advocacy skills permanently into food – something that affects everyone, not just the relatively elite, which postgraduates are, educationally speaking, in spite of their diverse backgrounds (only about 6 per cent of the population in Australia has a postgraduate degree).

#IMMABEAFARMER – IF YOU HASHTAG IT, IT WILL COME

Backtrack to the southern cliffs of Wales, August 1991. Stuart and I have met in London and been backpacking together for about three weeks. As we walked the cliffs, we fell deep into the kind of conversations common to young lovers the world over – who *arrrre* you and who do you want to be?

Me: I want to have a farm in Colorado, near Boulder where all the hippies are, and I'll have all the animals and all the artists and writers and musicians can come and stay for free and be inspired and we'll all read and write and play music together and grow everything ourselves.

Stuart: I can't see you doing that.

Me: <pouts> you clearly don't know me well enough.

What I thought I would do with all those animals given I was a vegetarian at the time, I'm not sure, but the point was I was a dreamer and my dreams were on the land.

Fast-forward to the end of Melbourne Cup weekend 2009 (don't get me started on its abuse of horses and gross displays of wealth and privilege), and you find Stuart, the kids and me heading home from a glorious weekend at a cousin's property in country New South Wales, and making a resolution to start the hunt for a property of our own. No firm plan, but we knew it was time to make the move out of the city and start growing all our own food at last. Serious searching commenced, but we had no idea what we were looking for, nor how we would make a living when we found what we wanted.

Eighteen months on and many life-chronicling blog posts later, we found ourselves at the Lake House in Daylesford in our favourite region in Victoria to hear Joel Salatin of Polyface Farm speak about building a local food system and how to 'scale up without losing your soul'. Of all the life-changing moments, this one was the most profound for both of us.

We'd never seriously entertained the notion of becoming farmers. Everybody knows it's impossible to make any money and you never stop working, right? But Joel countered that received wisdom, and assured us that not only can you make an honest living as a small-scale farmer, you can help change our food systems for the better while you do it, and, baby, it's a beautiful life, hard work included.

Joel foregrounded the need to farm in a way that respects 'the pigness of the pig', so he had me from the beginning. He also advocates for (and practises) supply-chain control, direct sales and farming within the ecological carrying capacity of your land (your farm should be able to 'metabolise its own waste'). We walked away from the seminar with the certainty that the way we would make our move to the country would be to farm, and, in particular, to raise pigs on paddocks. A new hashtag was born: #immabeafarmer.

It was immediately apparent to me that this was the move towards deeper knowledge of production practices that I'd been looking for – the farm would be the platform from which to advocate more strongly about the need for a revolution in both production and consumption practices in the Global North. My research had recently been flirting with production as I tried to better understand why animals were being raised in sheds, and plants in massive monocropping systems – what were the real barriers to ethical, regenerative production as I understood it?

Cosmopolitan research, as it turns out, finds that cosmopolitan subjects typically hold a strong environmental ethic as well as cultural ethic – they're more open to and engaged with the world. It was this cosmopolitan sustainability research rabbit hole that had led me to deeper considerations of production, which had led me to the openness with which I received Joel's 'you can farm' message. Yep, we could farm!

Another year of searching for properties, now informed by our decision to farm pigs, led us to what is now Jonai Farms & Meatsmiths. We found a reasonably-priced-for-our-region 69 acres on lush volcanic soils (the property actually largely comprises the southern slope of an old volcano), with good internal and boundary fencing, five dams, two large sheds and a livable home.

My writing on free-range pork and poultry went up a notch on the blog as I avidly researched everything I could find about these systems. I became increasingly disenchanted with free-range poultry systems, where, to put it as a recent ACCC finding against a so-called free-range grower did, 'people have a reasonable expectation that free-range birds will have more than the *potential* to go outside'. And yet many of the larger 'free-range' poultry farms do not give their birds constant access to outdoors, and some have stocking densities much higher than what anyone would consider free range – their birds may only have the equivalent of an A4 piece of paper to themselves and, in many cases, no fresh grass in sight, just dirt.

We visited free-range pig farmers and were blessed to make Anthony and Amanda Kumnick of Greenvale Farm our mentors and friends as we joined them in farming pigs ourselves. We read everything we could get our hands on about pig farming specifically and regenerative farming more broadly. (Note: the pre-1950 texts are best as they're generally predicated on small holdings and feeding animals with agricultural surplus or 'waste' produce rather than purpose-grown feed, and they assume your animals are outdoors, not managed in sheds with more issues about ventilation than soil.)

In a slightly alarming turn of events, before we'd bought a property, we were evicted from our suburban home by a landlord keen to renovate and increase the rent. The decision had to be made, and we finally made an offer on our beautiful land and took the plunge.

We then took a four-month detour to drive across America in a 1977 GMC motorhome (the RockVan!), visiting a range of farmers already doing what we hoped to do back in Australia. We started with a mandatory pilgrimage to Polyface in Virginia, where we got to see Joel Salatin's radically transparent system in all its glory. The man practises what he preaches – his animals are on the paddocks, the paddocks are lush from his multi-species soil regeneration system, and, along with a team of interns, he and his family slaughter and process all of their chickens on the farm.

Nobody blinked an eye when we rolled up the driveway and asked if we could poke around the farm; they just pointed us towards the pigs off in the forest. And when we met Joel at the end of our wander, he spoke fondly of his time in Daylesford and posed for a picture with me and the kids before we headed down the Shenandoah Valley to fry up some Polyface chicken and devour it with some local artichokes. This is what we were signed up for – a life of growing your ethics and enjoying the delicious results as the Salatins do.

One criticism I would offer of Polyface and its disciple systems in the United States and Australia is in regards to the meat chickens. The breeds used are (typically) fast growing and basically unable to walk well before slaughtering age at about eight weeks. Some will call them lazy, but from all appearances the issue is that the breed grows so quickly and, in the United States at least, the breasts are so large and the thighs so slim that the chickens can barely carry their own weight, so they mostly step, slump and wait for feed. Personally, I reckon that raising breeds that can't express the 'chickenness of the chicken' due to these traits goes against our ethical/ecological/regenerative impulses. That's letting the commercial imperative win, and it's back to the top of the slippery slope down to industrial agriculture.

We arrived at Jonai Farms on 1 September 2011, wide eyed at our bounty. After two decades of farm-like backyards full of herbs, vegies, fruit trees and the ever-present chooks in their Stuart-built dome tractor, we couldn't believe the scale on which we could now grow food. Not to mention the enormous responsibility that comes with a commitment to growing it commercially – especially a higher risk food like fresh meat. We gave ourselves five years to turn our first profit, intending that I would work off-farm as needed for extra income.

The first year lives on in our memories as a halcyon glow of basically homesteading. We got our first rare-breed Large Black pigs – a boar and five gilts (female pigs before their first litter); our first herd of dear little Lowline cattle; a flock of rare-breed Lincoln sheep; and a very motley flock of common and rare-breed chooks – from the terrible layers but gorgeous Speckled Sussex hens, Henrietta and Gertrude, to the best broody mama Pekin Bantams, Fern and Blacky, to the racy Anconas, bossy Auracanas, and reliable old common Isa Browns.

We'd find time to climb the volcano and appreciate the view of Mount Franklin, and walking the perimeter was something we did for fun and exercise, not to find a fault in the electrics. What we didn't do enough of was build more fencing in time for the population explosion that was soon to hit when those five gilts started producing litters of 8 to 12 piglets at a time, and we rapidly went from 6 pigs to 56.

I continued to write a lot on the blog about ethical production, and I was quickly gaining a sense that there was a lot more to it than I realised. And as well as writing on the blog, I was doing a lot of listening – feedback from ethical eaters, other farmers, and a disproportionate number of vegan abolitionists. People had followed our journey up to the farm via the blog, and we had orders for pork before we had pigs! Twitter was another fabulous point of connection with people of like and different minds on issues of production – a diverse community with which to test ideas and assumptions (and ask a lot of questions!).

I remember learning we needed a PIC – a property identification code – that goes on the paperwork required to be submitted to various regulating authorities every time an animal moves from one property to another, or goes to the abattoir. As I asked Anthony Kumnick for some advice on procuring this magical number, I queried if meat is so traceable back to every property the animal has ever been on then why can't consumers in the supermarket have that information in front of them? Of course, I now know that the traceability is not actually as tight as it seems when you release your animals to an abattoir – by the time that beef is in a butcher's shop, the butcher typically doesn't know where it came from beyond the abattoir itself. But theoretically they could, and so could you.

There's a contemporary misconception that epicureanism is just about being some kind of gourmand – someone who relishes and (over) indulges in food, drink and the arts. Hedonism suffers the same reputation, but in fact both philosophies are about the pursuit of pleasure with justice – 'neither to harm nor be harmed'.

Given the increasing interest in discussions of the ethics of food and agriculture systems, I think 'ethicureanism' captures the intent best for modern society – an ethical pursuit of pleasure, not just for oneself but for all. To take one's pleasure without taking it away from others. You know, be kind to one another.

Here at Jonai Farms we mindfully practise ethicureanism – we seek a pleasurable life for ourselves, our animals and our communities. Articulating that philosophy helps guide our every decision, and leads us to start from a principle of minimal intervention – the precautionary principle, really.

To give a practical example: we didn't first ask, 'How will we castrate the boars?' we asked, 'Should we castrate the boars?' On balance, we decided after a year that we would castrate as the boars were causing unwanted teen pregnancies in the herd, which aren't good for young gilts or their litters, and many of them had boar taint in the meat, which takes from the pleasure of those who eat the pork. One brief surgical procedure for the males eliminated a number of other instances where their virility took pleasure from others.

In the second year of farming, ready to sell our first meat, we encountered the challenge thousands of small livestock producers before us have faced – where and how to have animals slaughtered, butchered, packed and delivered. The abattoir was actually the easier end of that question as we only had one choice closer than three hours away. It's a big abattoir that primarily services the intensive piggery industry, which was definitely somewhat off-putting. However, they use carbon-dioxide chambers for stunning, which is globally recognised as the highest welfare option for pigs. And it's just over an hour from us, meaning we're not transporting animals a long distance.

The crazy situation Australia has ended up in, in just the past 20 years, is the closure of so many regional abattoirs, which means we grow animals in the country but most farmers must drive them to the city fringe for slaughter (and then farmers must listen to people worrying about the stress of transport, when no farmer I know would like anything better than a shorter trip to the abattoir).

And while we would all love a mobile abattoir option to do on-farm kills, current regulations prohibit any slaughtering facility for commercial meat from being mobile. There are farmers across Australia working on solutions of this sort, and in the United States there are mobile abattoirs that some describe better as 'portable', as they don't actually move very often due to the water input and waste output infrastructure required, as well as staffing, minimum number of animals to make a visit viable, planning schemes, etc. But know that many of us are working towards a solution of this sort, where animals really need know only 'one bad day' and never know it's coming.

Butchering, it turns out, was not as easily resolved as slaughtering in our case. The first butcher we approached basically laughed us out of the shop, the second asked, 'Why would I want to cut up your pigs?' (um, because we're offering to pay you?) and the third simply didn't have enough time to add a regular extra two pigs from us in his already thriving six-day-a-week retail business. And so we formulated a plan to build a boning room right here on the farm.

People told us we were crazy, that the regulators would never allow it, that it was too expensive ... and, typically of us, we ignored them and pressed on. We launched a crowdfunding campaign on the Australian online platform Pozible aiming to raise $21,450 in 40 days. We hit our target on day 19 and ultimately raised $27,570 – the first farm in Australia to crowdfund major infrastructure. We were frankly gobsmacked at the result – we were even supported by vegetarians, the best testimonial of all!

More than anything, though, we took our success as a clear sign that the tide really is turning and people are serious about wanting to know (and be involved in) where their food comes from and how it's produced, especially animals. And subsequent campaigns have confirmed this, such as the following:

* Bundarra Berkshires raised $18,135 to build a commercial kitchen and curing chamber for their free-range pig farm

* Madeleine's Eggs raised $67,000 for an egg cleaning and sorting machine

* Mt Beckworth Free Range raised $44,297 for their on-farm butchery.

There are more all the time as farmers turn to crowdfunding rather than the banks in a bid to feed our communities. Connectedness to community – aided greatly by today's social media tools – is critical to the future of Fair Food farming.

Fortuitously, right as we prepared to launch our campaign, we did find a butcher who would butcher our pigs fortnightly, and, even better, who was just crazy enough to take me on as an apprentice for six months while we built our own boning room. Sal (of Salvatore Regional Butcher in Ballan) is a dreamer in his own right, and loves a bit of gumption and willingness to do the hard yards. And so after-hours once a fortnight, he and I would transform our pigs into the normal range of cuts. I learnt to use the bandsaw first on shoulder roasts before progressing to the scarier work to cut chops, after being convinced that my naive old-school insistence that I would only use a hand saw and cleaver was wildly impractical.

I took to butchery with a passion born of 20 years of serious cooking and caring deeply about the provenance of my food, especially the animals. After a decade in front of a computer, I was surprised to find how deeply I enjoyed the manual labour of it. Just as Stuart needed to get out of high-rise offices with no opening windows and into the fresh air to pursue 101 projects at a time (this man was born to be a farmer), so did I need to do tangible work with my hands, not just my brains. Intense curiosity, obsessive DIY-ism and strong respect for the life taken for food combine to make me one mindful meatsmith.

And so, on 21 January 2014, two and a quarter years after donning my workboots as a freshly minted farmer, I stepped into our newly licensed boning room and took up my role as Mistress Meatsmith. I felt competent on the pigs, shaky on the beef, but confident that I could pull off this new chapter in our journey. When we took over butchering here on the farm, we immediately moved from running a small loss to turning our first profit.

I'm keenly conscious of allegations of elitism in producing food in the manner in which we do, which costs more than industrially produced food. How are the most financially disadvantaged in our society ever to access bacon that costs $32 a kilogram when they're spending over 50 per cent of their income on housing costs? Is it fair that those with the lowest incomes should only have access to highly processed and nutrient-poor foods rather than the ethical, nutrient-dense food produced by farmers like us? Of course it's not.

But driving farmers off the land, and caging pigs and chickens so one person can manage a shed with thousands of animals, is surely not the answer either. It's not fair to the animals, to the workers, to the environment nor to the eaters who can access that $5-a-kilogram bacon. We won't solve poverty with industrial agriculture – in fact, exploitation of labour is one of the pillars of that system.

We can rely for now on the excellent work of organisations like SecondBite, Foodbank, and Spade & Barrow to make nutritious whole foods available to those unable to access them otherwise, but these are just bandaids on a weeping wound that is systemic poverty and a broken food system. We produce enough food in the world already to feed 10 billion people, yet a failure of governance and distribution means too many go hungry.

One thing those of us in the movement can do is promote the principles of Slow Meat – eat better, eat less. People can afford to pay more for meat if they eat less of it. And many of our customers do just that – they can't afford our meat frequently, but only buy ours and other ethical meats as they live by slow meat principles.

The other thing we can do is find ways to make our produce accessible without devaluing its real worth. We regularly discount meat in the freezer even though it's still of high quality and are delighted by the strong response from low-income members of our community. Again, this is really a stop-gap measure that doesn't address structural poverty, but it's an intervention available to those committed to social justice as we continue our work towards a more just society generally.

I won't delve further into the real cost of cheap food – industrial agriculture externalises costs to the environment, to the health and wellbeing of workers and communities, and of course to the animals – and that's what makes it cheap in dollars only.

We regularly have veterinary students come for placements on the farm – most of them desperate to do a stint on a free-range farm as they can't stand being in the toxic air of the intensive piggeries. We teach them things that every free-range farmer knows but that they've been led to believe is, if not impossible, at least not a viable production model.

They marvel at our pigs' curly tails as they've learnt that pigs' tails must be docked or they'll chew them off each other due to lack of 'environmental enrichment'. That is, pigs in sheds will literally chew each others' arses off out of sheer boredom. Many vet students are astonished to see our sows farrow (give birth) outside, with only around a 15 per cent mortality rate from squashing and stillbirths. Most have been taught that free-range systems average up to 50 per cent losses from squashing, a total fallacy. They've learnt it's too cold in winter for pigs to remain outside. When they tell me this, I blink and say, 'Life isn't a fairy tale, you know. Pigs can't actually build their own houses, so what would you have them do?'

Ethical small-scale farming

Our farm was inspired by the philosophy of Joel Salatin's Polyface Farm, and we have a profitable business with the following elements:

- supply-chain control: we do all our own butchering and smallgoods and charcuterie manufacturing
- direct sales: we deliver our meat monthly to hubs around Melbourne and our region for eaters to collect, plus a few restaurants and through our farm gate shop

• radical transparency: we're open Monday to Saturday, from 10am to 4pm – all are welcome to tour the farm and meet the happy piggehs. We consider ourselves certified by the community

• staying within our ecological carrying capacity: we determined our herd size based on what's best for our soils in tandem with what makes us financially viable

• respecting 'the pigness of the pig': our pigs are free to express their natural behaviours of rooting, wallowing and running freely around the paddocks.

These are the best ways to ensure we are fair to our soil, the animals in our care, ourselves as custodians and workers of the land, and those for whom we grow food.

In the 1960s there were over 50,000 pig farmers in Australia, but now there are fewer than 700 commercial producers. And yet we produce more pork than we did back then. More pork almost exclusively from pigs in confined systems that have detrimental effects on the animals, the workers, the environment and the moral fibre of all of us who condone these systems with every bit of industrial bacon we buy. And, shockingly, more than 70 per cent of the bacon and ham consumed in Australia is manufactured from frozen, imported pork primarily from Denmark, the

United States and Canada, and then labelled 'Made in Australia'.

Since commencing this beautiful, hardworking life as an ethicurean farmer and mindful meatsmith of uncommonly delicious rare-breed pork and beef, I've gone from being a lone PhD student banging away on my blog about what's wrong with our food system to a deeply embedded practitioner activist in the Fair Food movement in Australia. Nearly every week I don my stripy socks, cowboy boots and butcher's fedora to climb another soapbox and make some noise about ethical, regenerative agriculture.

Along with some of Australia's leading thinkers on food systems, Nick Rose and Michael Croft, I was a founding member of Fair Food Farmers United (FFFU) – a producers' branch of AFSA – in early 2014, and at the end of 2014 I was elected president of AFSA. I hold a strong vision to support small farmers as they innovate and build a new system outside of the old one that's killing us, as Nick Rose has so grimly reminded us. If we support farmers, we support a whole new/old path of growing food in ways that are connected and transformational for everyone.

Ensuring all food is grown fairly means everyone has access to Fair Food. A total revolution is what we need.

We at Jonai Farms & Meatsmiths are just one of many leading the change, being the changes we wish to see in the world.

Watch out, Big Food, because we are legion, and the future of Fair Food is now.

PATHWAY TO AN UNDERGROUND INSURGENCY

CHARLES MASSY

Charlie Massy is the person in this book I've met the most recently – with the exception of David Pocock and Guy Grossi, who contributed the forewords – and the one I know least well. But as soon as we met – in February 2014, on the way to a food movement workshop at the Portsea Hotel, in Tammi Jonas's farm ute – I knew his story was one that every Australian needed to hear.

What Charlie has captured with his research goes to the very essence of this book: it reveals a transformation underway in Australia's food and farming system. And, as Charlie so clearly explains, the wider social transformation begins at the level of the individual. This is an enormously powerful and hopeful story, because its message is that we are not predestined to continue travelling down the same destructive path we have always trod. We have a choice, now and always, to take a different path.

I have Charlie Massy to thank for bringing this book to print, so that these stories of individual transformation – and this story of collective transformation – might reach a wider audience. I mentioned to Charlie as we drove away from Portsea Hotel on 28 February that I had an idea for a book, along these lines. He suggested that I should get in touch with his publisher, Alexandra Payne at UQP. Emboldened by his encouragement, I put my thoughts into words, and onto paper, and into an email. And, 16 months later, this dream of mine is finally seeing the light of day, thanks to Charlie.

There are wheels within wheels, you see. The criss-crossing paths woven between me and each of the 10 contributing authors, and among so many others, brought me and Charlie together that day in February 2014. And made it possible for you to read these words, and hear these stories, and feel their power and possibility. I'm beginning to feel there's no such thing as 'coincidence'...

***

Our ideas are part of the ecosystems we inhabit.

Lynn White

FROM THE AUSTRALIAN FARM TO THE MYANMAR VILLAGE

In May 2013, I was sitting under a tamarind tree in the central 'Dry Zone' of Myanmar, along with four members of the international aid organisation ADRA. Being tall and unaccustomed to sitting cross-legged for hours, I surreptitiously squirmed to adjust my aching legs. Facing us was a semicircle of Myanmar villagers, elegantly dressed in colourful sarong-like *longyis:* men in collarless shirts and women in modest, subtly patterned blouses. They sat comfortably squat-legged, males on the left, females on the right, in neat protocol. Behind the group of leaders was a curious but quietly attentive crowd, comprising the rest of the village of 100 or more people.

Their warm, friendly faces, and their generous offerings of sweets and fruit on mats on the ground, belied a quiet but deep desperation underneath. As we were painfully discovering, this was because their landscapes and society were coming apart at the seams.

The Dry Zone is an area of around 88,000 square kilometres (almost the same size as Tasmania) either side of the famed north–south flowing Ayeyarwady River in central Myanmar. It comprises 57 townships and hundreds of villages in the rural areas. The latter's members eke out a living from pastoralism and some crops, with a diminishing surplus available for trade and exchange. Though comprising but 13 per cent of Myanmar, the Dry Zone contains over 30 per cent of the nation's people: around 15 million.

As we drove across bumpy, often steep and eroded dirt roads to visit a number of villages over five days, signs of a desertifying landscape were everywhere.

Flop-eared black-and-white goats ranged freely, reduced to browsing on woody shrubs and thorn-bush, as all nutritious perennial and annual grasses, forbs and succulent shrubs had disappeared. Gullies everywhere were in active stages of erosion, and watercourses had become giant sand-rivers. The desolate landscape matched the quiet desperation in the welcoming, dignified people crowding around us.

All this is because the Dry Zone is at the forefront of global climate change. The village areas we visited receive a similar annual rainfall to my own farm on the Monaro in southern New South Wales: around 550 millimetres. The difference is that their rain is meant to occur during the monsoon period, from May to October. But over the last 20 years, the timing and quantity of monsoon rain has been severely disrupted. At the same time as temperatures have increased, total rainfall has decreased by up to 40 per cent, while the shortened monsoon rainy period frequently arrives either too early or too late, and is increasingly irregular. More frequently also, heavy rain events are often interspersed with long dry spells, causing early-sown crops to fail. As a consequence, the landscape is drying out and traditional grazing and cropping practices have not adjusted: hence the accelerated landscape degradation through over-grazing on their easily eroded soils.

No longer able to sustain their original village populations, increasing numbers of villagers (especially the youth) are leaving their traditionally secure village lives to seek employment in towns and cities elsewhere. Many end up in large refugee camps on the Thai border. When they return to visit their villages, such alien modern diseases as HIV and STDs are spread. Moreover, as all young males in this Buddhist society traditionally spend time in a monastery, such stays are becoming longer. This exposes them to increased sexual abuse: a totally suppressed issue in this society. The United Nations World Food Programme indicates that 42 per cent of households in the Dry Zone are now 'food insecure', and already there is evidence of malnutrition in the under-five-year-old cohort. The net result, due to climate change and an inability to adjust ancient cultural practices, is the degradation of both landscapes and society.

This means there are no resilience buffers left in the system. These communities are literally on the edge of a precipice, where a major shock – be it climatic, or a contagious animal or human disease, for example – could tip them over the edge. In many ways, therefore, this area and its beautiful people are the canary in the coalmine for much of humanity.

The question is: how did a 'farm boy' from a grazing property on the southern Monaro in Australia end up in central Myanmar seeking solutions to the desperate plight of these villagers? As a result of a long journey, the answer perhaps provides a pathway for how our own society (and, indeed, villagers in Myanmar's Dry Zone) can address the largely unacknowledged but dire issues of landscape and ecosystem degradation, and the corrosion of both the food we eat and thus human health. In short, the answers may provide solutions to humanity's ability to confront the new Anthropocene era.

A DIFFICULT APPRENTICESHIP

My mother died when I was four-and-a-half years old: another victim of a repressed but too common malaise in the 'bush' – suicide. As a self-sufficient only child who relished exploring the bush on our farm, or hunting rabbits among granite rocks with accompanying dogs and poddy lambs, I was packed off to boarding school at the age of eight. This small school of barely a hundred pupils largely proved a blessing, for it was surrounded by farmland and bush in the New South Wales Southern Highlands near Moss Vale. Besides companionship and sport, I could go camping and exploring in the surrounding farmlands and native forests. But the greatest blessing was a young teacher called John Hutchins who was a keen ornithologist, and who took under his 'wing' a small number of boys interested in nature. With nearby rainforests and tall wet sclerophyll forests on the coastal sandstone escarpment, weekends became a joyful discovery time in pursuit of elusive lyrebirds, rainforest pittas and other mysterious denizens of these rich ecosystems.

Consequently, on leaving secondary school and after two years jackarooing on the family farm under my father, I enrolled at the Australian National University in a science degree, with the intent to major in zoology. My heroes of the time were scientists like the founding ethologists (animal behaviourists) Niko Tinbergen and Konrad Lorenz. I wished to become an ethologist also, working in wild and beautiful regions of our planet.

However, the professor of zoology was a strict Oxford professor. One day I remember him ripping up an essay I had lovingly laboured over and which was the result of hours of trapping, colour-banding and then stalking and observing a curious, social-nesting bird: the white-winged chough (which was plentiful in the bush on our farm). 'What I want,' he said, 'is rigorous quantitative measurement, not this opinionated free-wheeling observation and supposition of yours!' To pass the unit (and ultimately zoology), I was required to count how many times a rat licked sugar off a lever as a reward in what was classic psychologist–behaviourist methodology.

The professor's view of reductionist science was not for me, and I soon gravitated across the campus to the first course in holistic thinking in any university in Australia, that of human ecology. Here, many of my fellow students and teachers were decidedly 'green', some radically so. Sometimes I joined them in protest marches on campus and in the city, for it was a time of a rising green consciousness in the wake of the tragedy of Lake Pedder. These marches concerned slogan-led issues like 'Save the Whales' and 'Keep Uranium in the Ground', though one day I recall one of my fellow larrikin students brought to a march a memorable handmade slogan that read 'Keep the Whales in the Ground'.

Joining this new, challenging (to the university at least) holistic-based course was to be a life-changing event. However, reality then intervened. My father suffered a major heart attack and I made the decision at the age of 22 to return home, take over the farm and finish uni part-time. This began a 40-year land management journey, but the start was not auspicious. Within a week, with my father in hospital and as I grappled with running a property when I knew next to nothing about management, I went for a drive one morning around our farm. To my horror I found 40 dead Merino lambs in a paddock, and others clearly sick. The dreaded bloodsucking 'barber's pole' intestinal worm had struck. Only decades later would I come to understand that this problem, like most in industrial agriculture, was human-made and because of compromised ecological management.

So my apprenticeship was a hard one as I initially became indoctrinated into the interventionist, nature-dominating, exploitative and traditional approach of industrial agriculture. Concerning our ancient, co-evolved landscape systems, only decades later would I appreciate that my taking over management at the age of 22 was the equivalent of placing a six-year-old child in a Formula One racing car on a busy highway. The chances of damage – even disaster – were extremely high.

HUMANITY'S INHERITANCE

Following the last ice age at the end of the Pleistocene period from around 12,000 years ago, our planet entered a unique phase in its history. Called the Holocene, this period lasted until the late twentieth century. It was a crucial phase for humans, who thrived because the planet's operating environment suited our species and many others besides. During the Holocene, the balance of atmospheric gases like carbon dioxide, oxygen and nitrogen also suited plant growth. The upshot of this confluence of events around 10,000 years ago (and particularly in the Fertile Crescent region of Eurasia) was the domestication by humans of a few animal species (initially the dog, then sheep and goat) and soon after this some food plants (especially annual grains and pulses). Thus began the development of agriculture.

After the emergence of modern humans some 1 million or so years ago, and then their spread out of Africa 80,000 to 90,000 years ago, the cultural emergence of agriculture was *the* crucial development in the entire history of our species. This is because it would come to have monumental significance for most other life forms on earth and its safe operating systems.

First, domestication and the better management of food security, aligned to interventionist management of natural systems, led to settled human habitation. From this devolved urban living, and with it the rise of city-states; of entire civilisations and their religions and belief-systems; the unleashing of ongoing cycles of technological innovation; and eventually, from the sixteenth century on, the interconnected scientific, industrial and capitalist revolutions that culminated in the Enlightenment. All of this climaxed in the modern globalised, industrial-capitalist economic system.

Second, however, this extraordinary cascade of events unleashed by a nascent agriculture carried with it the seeds of potential destruction for the inventive human species. It was as if humanity had made a pact with the devil. In exchanging a hard hunter-gatherer existence that nevertheless held us close to the earth like other species, we humans took metaphorical wings so as to fly towards an easier, self-satisfying, more comfortable life. But the Faustian bargain is that now, like Icarus, we too closely approach the sun. Industrial agriculture and modern technology have enabled profligate living for the exclusive but minute minority of the rich; a miserable existence for the vast majority of the poor; a human population that is set to reach 9 billion in less than 35 years (a level around three times more than can currently be sustained by the earth); and a planet whose self-sustaining operating systems are now being corroded (in some cases possibly irreversibly).

Australia, being the last and most recently settled, conquered and dominated of the earth's habitable continents, is at the forefront of the fallout. The settlers who 'conquered' the Australian continent through the nineteenth and twentieth centuries brought the arrogance of the 'mechanical' mind, which saw nature as inanimate, passive and machine-like, and so available for exploitation. They saw the mysterious indigenous occupiers of the Great Southern Land in the same light, not realising these fellow humans possessed belief, knowledge and spiritual systems that held a template for long-term survival in this unique and fragile continent. This is because Indigenous Australians (like fellow humans of an 'organic' mind elsewhere across earth) possessed knowledge that had allowed survival for a hundred millennia and that stretched back more than ten times longer than the comfortable Holocene. The wisdom and collective experience of the 'organic' mind implicitly understood that we humans were part of an organic cosmos and nature, not separate from nor 'above' it; and that because nature was seen as a nurturing 'mother', we had to both care for her and live within her cycles and limits. This 'organic' mindset therefore carried strong ethical, spiritual and ecological implications.

We, the present generations, are the inheritors of humanity's monumental Faustian bargain. It is now increasingly acknowledged that planet earth has left the Holocene era and entered a new age, termed the Anthropocene. It is so named because this is the first time in the history of the earth that one species – *Homo sapiens* – has the potential to destroy most of life on earth, including our own species. And this is because our behaviour, in releasing ancient stored carbon from fossil fuels and other derivatives, has begun to destabilise, even threaten, the safe operating limits of Spaceship Earth. Many of the world's leading scientists believe we have already crossed (perhaps irreversibly) three of these safe operating limits: climate change, biodiversity loss and a balanced nitrogen cycle.

However, following a lifetime's management of an Australian landscape, I have emerged full of optimism that we humans can address the Anthropocene era, that having got ourselves into this mess, we also have the capacity to get ourselves out of it. This is because, while my management life was filled with mistakes, it was this that awoke me to the dangers of industrial agriculture. In turn, this led me to examine new developments in regenerative agriculture. In the process I have come to understand that industrial agriculture is strongly implicated as one of the major causal factors in humanity either crossing or threatening nine out of ten of the earth's safe operating limits. What is clear, therefore, is that a key pathway to address the issues of the Anthropocene era and turn things around is through agriculture. This links to the second leg of the pathway: a parallel galvanising of our urban sisters and brothers via healthy food and its connections.

That is the story I now want to relate: how I came to realise that we humans really do have the capacity to take ourselves and the earth out of the Anthropocene era into what the prescient philosopher Thomas Berry calls the Ecozoic. I believe agriculture is the pathway because of its capacity to regenerate healthy landscapes, healthy food and healthy people and societies. But we cannot make this transition unless we move from existing in the 'mechanical' mind to a 'neo-organic', earth-sympathetic mind.

BECOMING DISILLUSIONED WITH INDUSTRIAL FARMING

Not every sheep dog on a working farm makes the grade. Some never learn or exhibit their genetic potential, and necessarily have to be dispensed with. When I look back at my early farming career, I realise now that, were I a sheep dog, I would probably have ended up 'down the back creek', so slow was I in coming to understand the real issues and limitations involved in landscape management.

Thrown in the deep end aged 22, I widely sought advice: from the district's 'best' farmers, Department of Agriculture and CSIRO personnel, my father, relevant literature and farming books, and all at the same time as I attended every relevant 'field day' on offer. This resulted in me being inducted into an industrial farming paradigm. I came to believe I had become a 'good' farmer. Aside from some rotational grazing on monoculture lucerne paddocks, the rest of our 1820-hectare (4500-acre) property was 'set-stocked', where sheep and cattle were put in paddocks and left there for long periods. Every year I attempted to put in new pasture and winter forage crops. This involved me ploughing up the earth and inverting the living, healthy 'A' horizon soil under the upturned, dead 'B' horizon. I would then savagely attack with discs and harrows this denatured soil so as to 'control' and 'kill' the dreaded 'weeds'.

Sometimes I ploughed country too steep, only to be 'unlucky' as heavy storms gutted sections of the paddock, sending a thousand years of topsoil down the gully and creating mini crevasses across the paddock. But that was how it was done, and you couldn't help 'bad luck'. Frequently the new winter crop or pasture only half-succeeded in establishing – 'luck' running out because the latest grass-seed or legume species (courtesy of mainly US multinational chemical and agribusiness companies) perhaps hadn't acclimatised, or else (we were told) perhaps we hadn't put on the right amount of synthetic fertiliser or manufactured inoculant. Or else I had simply been 'unlucky' because the weather patterns had been 'abnormal'. My own rationalisations, like the excuses from agronomists and ilk, were endless.

Then came my first big five-year drought, from 1979 to 1983. Up to then my management methodology was based on the assumption of 'normal' seasons. But the reality is that severe 'dries' and droughts are the norm in Australia, and my mind was unprepared for this. So, as the 1980s drought worsened year by year, I attempted to fight it, perpetually waiting for the 'normal' season to return. Our land began to lose its grass cover because I 'had' to maintain our 'normal' stock numbers. Our debt grew as I bought more grain to defend these numbers and 'to fight' the drought. At the same time, and on the best advice, I thought that, to beat the cost-price squeeze, I needed more land and 'bigger scale'. So we bought a neighbouring block, further increasing the debt.

During this time also my wife, Fiona, and I had three young children whom we tried to raise on home-grown vegies, fruit and fresh milk from a Jersey cow I milked, while at the same time my father was dying. Long, hard days of feeding sheep, running a farm, constantly repairing or 'finding' water, and then lying restlessly, sleeplessly in bed, forever running financial, feed and stock calculations in my mind, all resulted in near exhaustion and, I am sure, an unrecognised mild depression. The irony is that at the time – and despite my love of nature and my attempts to grow healthy food for my family, plus a collections of books by leading organic farming innovators sitting on my office shelf – I did not see there could be a better way to approach broad-scale farming. Somehow, my mind was still locked deep in the industrial farming model, and this turned me blind towards any alternative. The double irony is that I later discovered the 1979 to 1983 drought caused other farmers (who at the time possessed far less affinity to the natural world than I did) to make a radical shift to a new, earth-regenerating agriculture. In retrospect, therefore, it was clear the key issue lay within my mind: with what a farmer later described to me as 'that one square foot of real estate between our ears'.

The power of a mind's blindness deeply inculcated in a paradigm was brought home to me years later when I reflected on my sheep-classing consultancy. I call this phenomena 'the vegie garden paradox'. On my travels across six states, I would turn up to many a farming family's homestead. Frequently, inside the fence surrounding the house and garden was a healthy household vegie garden: no chemicals used and natural manures or fertilisers applied. Then, after a cup of tea, we would leave the home garden for the sheep yards. But this was like crossing into another world, because outside the back garden gate I would discover an industrial landscape: large piles of chemical drums behind sheds; massive spraying and tillage machinery in the yard; and a landscape either 'nuked' by chemicals or carrying sheep that were tinted red from walking powdery-dry soil in flogged-out stubble paddocks. Conversely, a farmer would proudly take me for a drive to see his revegetated riparian corridor courtesy of Landcare or 'Catchment Management' money. But to access this corridor we would drive through similar flogged-out country.

What was going on?

As I farmed through the 1990s and then 2000s, I became increasingly unsettled about my own approach to farming. By the late 1980s I had begun developing a Merino stud that pioneered new thinking and application in molecular genetics and biology: a value-added genetic wool- and meat-marketing business would enhance my farm's viability. With the genetic business went an innovative approach to marketing and processing elite, customer-friendly wool fibres to international customers, during which I wrote the first world protocol for processing 'organic' wool. But I learnt hard lessons about attempting innovation in deeply conservative and power-laden industries, and again about the depth and power of long-held paradigms. This was reinforced by a painful time on statutory and international wool boards, where I experienced the strength of naked power among large multinational companies. I also encountered powerful male egos defending their entrenched patches and industrial economic rationalist 'thinking' in Merino breeding, politics, science and processing.

I was a slow learner perhaps, but my concern about the impact of my management on a landscape I loved led me to investigate some new ecological grazing approaches. As early as the post-drought 1983 year, I had begun experimenting with ecologically based methodologies such as the Keyline approach, which spawned the worldwide Permaculture movement. But the results were, I thought, too slow coming and I undid the good work with traditional 'set-stocking' anyway. Clearly, I was not yet ready. But in the early 1990s I attended an ecological grazing course and enthusiastically adopted what was then called 'cell grazing', where large mobs of sheep and/or cattle are rapidly rotated around paddocks, with the aim of regenerating grassland ecosystems. However, again I didn't fully catch the ball, as a growing stud Merino and consultancy business meant I only partly, and therefore ineffectively, adopted this new thinking.

In the meantime, as I travelled more widely across Australia, I noticed two things. First was the ever-increasing degradation and deterioration of our landscapes: from the spreading 'white cancer' of salinity; to over-clearing of the Esperance sand plain, Queensland's brigalow country, or the rich native grasslands of New South Wales's northern black soils; and to the widespread and increasing intensive use of herbicides and pesticides across all farming landscapes.

But the second thing I noticed was that a number of our clients (already sympathetic to adopting new ideas in sheep breeding) were beginning to either experiment with, or adopt, some of the new forms of ecological agriculture.

Eventually, burdened by debt and overwork plus an ever-deepening disquiet about my own halfway approach to landscape management, I dispersed our Merino stud in 2007. By then we had fully swung over to holistic grazing management, and soon after I enrolled at the ANU's Fenner School for Environment and Society to do a PhD in human ecology.

What drove me back to university wasn't just my increasing disquiet with my own behaviour, nor my concern about what was happening to our farming and natural landscapes. No, the third reason was the clincher: that more and more I had begun encountering exciting, profitable and transforming nature-sympathetic farming approaches that were actually *regenerating* our landscapes.

So I set out to examine that 'one square foot of real estate between farmers' ears', both from my own perspective, and especially those of the farmers who had been able to make a transformative change from harmful and ecologically destructive industrial agriculture to positive, profitable, nature-regenerating landscape management.

TRANSFORMATIVE FARMERS

In 1967, in a famous paper in the journal *Science,* Lynn White stated that 'our ideas are part of the ecosystems we inhabit'. It wasn't until I interviewed 80 farmers across Australia as part of my PhD research that the truth in this statement became evident to me. Take, for example, one of the first holistic grazing managers I visited, Tim Wright on the New England Tablelands west of Uralla in New South Wales.

Unlike me, Wright was stimulated to profoundly change his management, thinking and worldview by the traumatic experience of the 1980s drought. By the early 1990s, when new thinking in the area of holistic grazing was brought to Australia by its developer, Zimbabwean ecologist Allan Savory, Wright was mentally ready. After attending a number of courses, he plunged in boots and all. This entailed combining sheep and cattle into mobs up to ten times larger than normal, subdividing country so he had more paddocks and then moving his big herds more rapidly: a process that mimicked the big migratory herds in Africa. Such natural massive animal migrations and movements had resulted in extraordinarily healthy grassland and soil ecosystems. The impact of mass hooves in rapid migratory movement, the recycling of urine and dung, and long rest and recovery periods for a diversity of plant species (and particularly deep-rooted perennials) led to richly diverse, healthily functioning and cycling ecosystems that were in balance at all levels: from pest-controlling parasitic insects and birds, to deeply functioning soils, and across the entire food chain, both above and below the ground. It was this model that Allan Savory has since adapted to grassland ecosystem management across all the world's major continents, and which Tim Wright plunged into, becoming one of the first adopters in Australia.

When I visited Wright in 2011 he had been managing his new system for 17 years, and the results were truly transformative: ecologically, financially and personally. Much of the ecological change he has triggered has been studied and collated by scientists at the nearby University of New England since he made his management shift, and corroborates the startling results in the paddocks and on the 'bottom line'. Ecologically harmful and costly industrial fertilisers, chemicals and invasive tillage technology and machinery had been dispensed with, which also led to huge savings in time and money. Yet Wright's soils were healthier in terms of the mass of microbial and other life, their function, their ever-increasing depth and their increased levels of soil carbon, along with increases in available nutrients like phosphorus.

Compared with traditionally managed neighbouring farms, his farm's resilience to droughts and extended 'dries' had vastly increased; the water courses had positively transformed; and his entire land had begun regenerating in spectacular fashion, with huge increases in plant, insect and animal diversity. In the process of constantly dividing paddocks to create more grass for his ever-rapidly rotating herds, Wright had taken his stocking rate and thus profitability from 8000 sheep units to a conservative 20,000. He now found life stimulating and exciting, his environment full of new surprises, and the ecological process one of ongoing and open-ended change and regeneration. Lynn White was correct: our ideas and worldviews are *the* key driver in how our landscapes function and how healthy they are. Wright and an ever-increasing cohort of peers are living proof of this.

At the end of my visit Wright summed up his journey. 'The thing that I find so frustrating talking to people is that to me it's common sense what we're doing,' he says. 'It's obvious that you work with your environment.' And this, he says, is because, 'we're managing more than a farm. We're managing ecology; we're harnessing sunlight energy through grass; we're managing soil and re-hydrating the landscape. We're managing biodiversity and economics, and' he concludes, 'we're also educating other people in the world.'

THE HARMFUL COSTS OF INDUSTRIAL AGRICULTURE

It would have been a relatively simple step some ten millennia or so ago for humans to manage the first domesticated and compliant sheep and goats on the abundant, rich and diverse Eurasian grass-lands. However, the domestication of plants – and particularly annual cereal and pulse varieties – soon after involved a different and more complex step entirely: one that, in time, would have enormous consequences for the global environment.

In any healthy grassland ecosystem there is a diverse mix not just of perennial and annual grasses but also forbs and often shrubs and trees. All work together in a self-sustaining, balanced and resilient fashion. It is this trait, in combination with a diversity of other species and plant types, that is the basis of Savory's holistic grazing management system, which Tim Wright was using.

Nevertheless, the key to humankind's cultural evolution and eventual massive explosion in population was to cultivate annual plants with mass seed production and soil disturbance. In time, via clever selection, plants like wheat, barley, oats, millet and rice were bred for ever-increasing production. With this went evolving techniques to control weed and perennial grass competition, and irrigation to boost production in dry times. But both these management techniques were high risk to a sustained and long-lived functioning soil ecology. Excess watering and resultant salinity or loss of aquifers proved the undoing of many a civilisation through history. Incorrect tillage systems did likewise. This is because to grow crops one must kill other vegetation, irrespective of its benefits. Intensive tillage (like ploughing and cultivation) to control weeds and perennial competitors destroyed soil structure and caused erosion and the leaching of nutrients. This in turn required humans to add more nutrients (initially in the form of animal manures or 'green manures' from other annual crops, and later in manufactured fertilisers) as compensation for the destruction of natural healthy soil function that otherwise fed and sustained both perennial and annual grasses.

With the rise, and particularly the intensification, of industrial agriculture from the early twentieth century, the artificial management and resultant degradation of nature's system vastly escalated. The massive use in the 1980s of biologically harmful synthetic fertilisers such as superphosphate, plus enormous doses of pesticides and weedicides, was followed by an increasing use of chemically dependent, genetically modified (GM) crops. Mass chemical use has thus emerged as an ever more dangerous and damaging substitute for old cultivation techniques.

Being vastly different and more fragile than the moister, younger and richer soil environments of Europe and the United States, where industrial agricultural techniques were evolved and refined, Australia's ecosystems sustained (and continue to sustain) enormous damage from modern agricultural techniques. The destruction was epitomised by the 'dust bowls' that have regularly occurred since the late nineteenth century, when millions of tonnes of long-evolved topsoil regularly flew east onto the ocean or to turn New Zealand's mountains red. This ecological damage has only escalated since, the spreading 'white cancer' of salinity being but one of many symptoms of crashing ecosystems across tens of millions of hectares in Australia.

What so excited me about my visit to Tim Wright and similar farmers (and led to my own management changes) was that the world now had an ever refining, open-ended, landscape-regenerating and profitable management approach for grazing animals, grazing being the occupation that is the world's largest, and which covers the greatest amount of the planet's usable land (some 26 per cent) while also employing some 1.3 billion people. Such practices thus create livelihoods for 1 billion of the world's poor. Crucially, livestock products also contribute around 30 per cent of the protein of human diets, at the very time that global fish stocks are crashing.

However, much of the world's food also comes from cropping annual plant species, which covers a smaller percentage of the planet but is far more intensive and interventionist. As seen above, this approach increasingly elicits huge, and ultimately unsustainable, environmental and human health costs as a consequence. In the face of the high chemical, high fertiliser, big machinery and GM onslaught by multinational interests and their dependent farmers on the planet's fragile and crashing ecosystems, until the early 1990s there seemed to be nothing radically new on the horizon that could provide a win-win ecological solution to our broad-scale annual cropping and food systems and the dire and increasing need to feed humanity.

In effect, the basic principles of producing much of our food via cropping had changed little over the last 9000 years or so, except to escalate technologies to higher levels of harm. However, and in what is proving a profound breakthrough, a revolutionary new approach has emerged out of some dusty paddocks in the central west of New South Wales. This approach is based on working with, and not against, Mother Nature; some humble farmers are its innovative pioneers; and it goes under two main appellations: 'pasture-cropping' and 'no-kill cropping'. Some of its practitioners also call this approach 'farming without farming'.

FARMING WITHOUT FARMING

Having had my mind rattled and upended by Tim Wright, soon after I visited another farming innovator I knew near the old gold-mining town of Gulgong in the New South Wales central west. Colin Seis appears to be your archetypical laid-back, laconic Australian 'bushie'. His speech is full of colloquialisms, his open, sun-blotched face frequently creased by a smile that betrays his dry sense of humour. Until only 25 or so years ago this hitherto apparently 'typical' farmer, who was in love with his Merino sheep and Kelpie working-dog stud plus a regular beer at the pub with mates, was a traditional cropper and grazer. His long-held family farm had annually been mercilessly ploughed and tilled for crops over four generations. This had then been followed by ploughing and fertilising again to build a grazing system based on the wide-spread formula of annual sub-clover and introduced grasses; while his livestock were set-stocked or then grazed on exposed earth in stubble paddocks. But all this began to change overnight, when in January 1979 the perennial spectre of an

Australian bushfire descended on his farm. Most of his fences, livestock, farm buildings, woolshed and house were destroyed, and Seis was left with virtually nothing. 'I was effectively stony broke,' he told me.

Severely burnt in the fire himself, Seis lay in hospital for three weeks, turning over in his mind how he could start again when he didn't even have the cash to plant a crop. But, as sometimes occurs, out of extreme adversity, combined with a confluence of chance events and some luck, came an innovative agricultural break-through that can be regarded as a major step in agricultural history.

The initial breakthrough came from extended discussions with his fellow collaborating farmer, neighbour and friend Darryl Cluff. The two regularly pondered Seis's cash dilemma but also the rising costs involved in modern cropping and pasture improvement and their associated ecological damage. Their discussions led them to pose the question: why not try putting in improved pasture at the normal time in autumn to early winter but without any costly fertiliser, especially since Seis couldn't afford it? Over the next few years both farmers experimented with this new minimalist approach. At the start, however, though saving on expensive fertiliser, they used cheap herbicides to control weeds and then direct-drilled in their introduced species. Initially, they found their introduced plant varieties – essentially drug addicts dependent on human-made inputs – rapidly died or yielded very little. This was aided by the early 1980s drought.

Then two things happened. First, native plant species (which are adapted to Australia's low phosphate soils and dryness) began to increasingly appear. These included valuable perennials and other species not seen for many decades. The second development was when Seis heard about the new holistic, ecologically based grazing systems being propagated by Allan Savory around the world from the late 1980s. Seis quickly experimented with the Savory system, but on his terms, telling me he set out to prove the new approach was just another false fad, 'a load of bullshit'. But, as he concluded, 'within about three or four months I'd bloody well proved myself wrong on that one'. Soon after this Seis did a holistic grazing course, and, like Tim Wright, began combining his sheep into large mobs, with rapid rotations and long rests. As with Wright also, one of Seis's aims was 'to have 100 per cent ground cover 100 per cent of the time'.

The major breakthrough evolved by the mid 1990s. It arose from another social drinking session with Darryl Cluff. As Seis recounted, a few beers too many probably freed their minds, and they began brainstorming out-of-the-box ideas. The result was that in 1992 Seis first combined the two new systems in his cropping enterprise. Using minimum fertiliser and chemical inputs, he began growing crops by direct-drilling cereals into native pasture paddocks that were dominated by winter dormant perennial species. 'I used to spray them after the first frost when they went dormant anyway,' he said. 'So I just chucked away the chemicals.' With his dormant grassland still intact, Seis then rotated his large mobs of animals so as to graze the edible winter cereal crops that had by now grown through the grasses. From the very first year Seis noticed some startlingly positive results, and by 1995 he had swung his entire cropping program over to what he now began calling pasture-cropping. In the years that followed, Seis constantly experimented with and fine-tuned his system.

Year by year, Seis found the new system got better and better. Despite dramatically slashing fertiliser, chemical and fuel use (and thus costs), Seis's cropping yields over the long term are equivalent to those of farmers using traditional agricultural techniques in his district. But in addition he has experienced similar financial and ecological benefits to Tim Wright over his entire farming and grazing system. He is running more animals more easily and cheaply, and his grassland and landscape have much greater diversity (plants, insects, soil biodiversity). He has no soil erosion, patches of salinity are disappearing, insect infestation and damage to his crops is now negligible, and his country is more resilient in terms of withstanding dry times. In short, by working with and not against nature he is experiencing multiple wins. Like Wright also, Seis regards 'weeds' differently; he sees most as beneficial agents in positive ecological gain. In this he is following in the footsteps of many great organic thinkers: not the least of them that famous philosopher Eeyore, who, in AA Milne's *Winnie-the-Pooh,* observed that 'weeds are flowers too, once you get to know them'.

Seis now grows a variety of crops with his pasture-cropping approach. He has also successfully grown vegetable crops using the new methods, and he is now refining the next stage in his innovative system: what he calls 'multi-species pasture-cropping'. He also talks about 'vertically stacking enterprises' to build resilience into farming. As part of this he sells up to eight varieties of valuable native pasture seed from a harvester he specially adapted.

As often happens in an innovation cycle, only two hours away, near Narromine, fellow farmer Bruce Maynard had independently come up with a similar combined direct-drilling, native pasture-cropping and holistic grazing system from 1995. However, because he used no chemicals or fertilisers at all, while also using a different, minimum-disturbance disc technology, Maynard labelled his equally profound innovation no-kill cropping.

The triumvirate of Cluff, Seis and Maynard has initiated one of the truly significant advances in broad-scale agriculture seen since domestication many millennia ago. What they have delivered is a universal approach to growing food and other crops that combines cropping and grazing, where each benefits the other synergistically and symbiotically: both ecologically and financially. This ecological farming approach has now spread to Europe and North and South America, and Seis and Maynard are committed international speakers and educators on their topics. Both are also keen self-educators, reading widely and interacting with many scientists, and each thinks and speaks like farm-cum-systems ecologists.

Echoing the views of leading regenerative agriculturalists and innovators I interviewed who had enacted a complete personal and farming transformation, Seis told me that 'we need to get back to natural design and enhancing ecological function. In fact, what we're doing here is a sort of Aboriginal land management. We work with nature.'

Principles for good farming

In 1993, environmental historian Donald Worcester described three principles for 'good farming':

- it should make people healthier
- it should promote a just society
- it should preserve the earth and its network of life.

After years of researching and practising regenerative agriculture, I would add a fourth principle:

- it should regenerate, not just 'sustain' or 'preserve', the landscapes in which we farm.

As I continued my journey across Australia, I found the same. I discovered farmers who were no-kill and pasture-cropping in alleyways between forage shrubs such as tree lucerne and salt bushes, and who had effectively droughtproofed themselves. Their soils were not just regenerating, they were also discovering positive and unexpected synergies with increased bird and insect biodiversity and vast reduction in insect damage to crops, while observing complex ecological interactions occurring between the grassland and the shrubs.

Likewise, farmers who had focused on revegetating their farms to ecological or Permaculture design were experiencing remarkable ecological, production and economic benefits that far outweighed the land taken out of production. And this was repeated across different case studies and different farmers that I studied: whether involving those focusing on biological or organic agriculture, biodynamics, re-hydrating landscapes via water retention and management techniques, regenerating native grasslands, integrated agroforesty, and even farmers overlaying the above approaches by channelling 'earth', 'cosmic' and water energies (what are called 'subtle energies') across their farms.

Without exception, in every case study and with every farmer these transformative practices ultimately came back to one key factor: regenerating and developing healthy soil. Degrade the soil and we degrade both ourselves and the very planet that sustains us. In contemplating Lynn White's observation that our ideas are part of the ecosystems we inhabit, it is clear that the present thinking of our modern society and its leaders and consumers is not far short of suicidal. We have entered the Anthropocene yet we ignore the warnings. Business leaders and their lackey political leaders, along with their collective henchmen in puppet organisations such as the International Monetary Fund, World Bank and World Trade Organization, seem to exist in a parallel universe, as if we can continue living, exploiting and plundering in a life separate from our sustaining ecosystems and a rapidly ailing planet.

Albert Einstein and many great thinkers since have repeatedly made the simple observation that 'today's problems cannot be solved with today's mind'. In other words, in Australian bush parlance, 'the thinking that got us in "the shit" won't get us out of it'. Thankfully, there are groups of people leading the way with a new 'mind' and new thinking. At the forefront are not just these leading regenerative agriculturalists and like-minded environmentalists, but also a huge cohort of urban-based people looking to regenerate their living environment and its patterns, and above all to reconnect with the healthy food we were designed for and that can only come from healthy soils and landscapes. This is our hope for the future. And that is why Donald Worcester's first two precepts for 'good farming' – that it should make people healthier while promoting a just society – are extraordinarily pertinent.

This latter issue of equity and fairness derives from the universal right of people to be able to freely access and/or produce healthy food without compromising their health or the ability of future generations to do the same. And that is where the exciting new movements across the world striving for food and societal systems that include fairness, equity and access to healthy food are so important.

THE NEO-ORGANIC MIND

Leading environmental thinker Gus Speth in 2008 observed that 'a system that cannot deliver the wellbeing of people and nature is in deep trouble'. As a consequence, he said, such a system 'invites ideas and actions that are transformative'. One of the regenerative farmers I studied put it another way: 'We have to change our mindscapes before we change our landscapes.'

One of the key findings of my PhD research and studying farmers who have fully committed to regenerative agriculture is that to enact a complete shift from the 'mechanical mind' and industrial agriculture, they have undergone a comprehensive shift in their thinking and feeling. They have experienced a personal transformation. In fact, among the leading innovators and practising regenerative agriculturalists, such a farming shift was not possible without personal transformation.

As with transformative learning in other areas, 60 per cent of the farmers I studied who had made this complete shift had changed because of one or more major shocks in their life. For Tim Wright it was the traumatic 1980s drought; for Colin Seis it was this drought and a bushfire; and for others it was a range of factors, from divorce to being poisoned by chemicals. Such shocks in turn seemed to crack the eggshells of their mind, laying them open to new thinking, new feeling, new imagining. Of the remaining 40 per cent or so who shifted, change was instigated by a series of small destabilising shocks.

What was striking about those who had made the change was that their very language changed. In contrast to industrial agriculturalists embedded in the nature-dominating 'mechanical mind', regenerative agriculturalists used language that was far more feminine, inclusive and empathic, and was full of holistic and ecological terms. They also exhibited strong emotional and ethical–moral components regarding their farming and living. They saw themselves as only a small part of nature, and they felt it was imperative they work within, and strive to enhance, her systems.

As I touched on earlier, the big change in human society from the sixteenth century or so had been from the ‘organic mind’ to the ‘mechanical mind’. My research made it clear that regenerative agriculturalists had enacted a major new shift to what I call the ‘neo-organic mind’. This involves a return to the nature-empathic view of the ‘organic mind’ but also is combined with modern scientific and systems ecology knowledge.

In other words, such transformative agriculturalists, and their urban cousins who have made a similar shift in their approach to food systems and living, have given us a template for what is needed to turn around the Anthropocene. Aldo Leopold once said we need ‘to think like a mountain’. I believe that if we are to save our planet, its safe operating systems and the gorgeous diversity of life residing on it, then we as a species need to overthrow the ‘mechanical mind’ and shift to the ‘neo-organic’. This can only be done by a life-changing personal transformation, akin to learning to ‘think like a mountain’ or ‘like the landscapes we manage’. This will allow us to rediscover our true selves: as, when everything is said and done, we are the product of aeons of evolution of our natural universe, its mountains, ecosystems and other life, and not the creation of a ‘mechanical mind’.

THE POWER OF STORYTELLING

The challenge now is how to rapidly accelerate a multiplicity of personal transformations. A series of inevitable looming shocks will unfortunately help in cracking the eggshells of our minds. But on the positive side, I believe it will also, and must, be the power of *story.* We humans are made for story. It is 'story' that goes deep into our hearts and minds. The current despair and ennui of our listless, sick and rapacious societies is because we have no story of integrity that fires our hearts and minds, that galvanises us into righteous, moral, ethical and rescuing action for ourselves, our families, our societies, and, wider still, our ecosystems and planet.

Well, we now have a great story, the greatest there could ever be: that of saving ourselves, the planet and all species and ecosystems on it. This story now needs to be told in diverse ways so it can help trigger the grand transformation of our society from the mechanical mind to the neo-organic. David Korten in *The Great Turning* states that such a turning 'requires reframing the cultural stories by which we define our human nature, purpose and possibilities'. What Thomas Berry calls our 'Great Work', or this cultural 'turning', involves, as he describes, our challenging 'attitudes that are so deeply bound into our basic cultural patterns that they seem to us an imperative of the very nature of our being'. Indeed, our next actions need to be our 'Greatest Work', and they must come from the 'Greatest Story' ever told.

This is why I am incredibly optimistic, because such a transformation can only come from the individual. Transformation won't come from the so-called great captains of industry and their giant transnational businesses; nor from their captive politicians and associated international monetary and trade organisations; nor from mainstream media, equally captive. Instead it will come – is coming – from individual farmers changing their landscapes and food quality; from the urban food consumer and the city, rooftop and suburban gardener changing their actions and decisions; and from a plethora of activists whose hearts and minds are already shifting to the neo-organic: as they change their local suburb, town, garden, creek and river, parkland, farm and catchment area. My friend Kerry Arabena (a professor of Indigenous health at Melbourne University and of Torres Strait Islander background) has reiterated that we all must now become Indigenous-like in returning our entire culture back to one based on solar, not fossil-fuel, energy. And this, she says, is because 'we are all indigenous to the universe'.

I call this movement, rooted deep at the very beginning of healthy soil and ecosystems, an underground insurgency. The power of the changed individual is truly unstoppable and uncontrollable, and that will enact a cascade of transformations culminating in the Great Turning. In his book *Here on Earth,* Tim Flannery states, 'While we humans may be built by our genes, our civilisations are built from ideas ... It's not so much our technology, but what we believe, that will determine our fate.' In other words it comes down to an inspiring story that will initiate our Great Work.

Recently, I received an email from my friend Brendon Irvine who leads the ADRA aid effort in Myanmar, and with whom I worked on my visit there. He informed me that they have now received funds to initiate a pilot program involving holistic grazing and other ecological approaches in three villages in the Central Dry Zone. The leaders of these villages understand, and are excited by, the potential of landscape regeneration via these methods. Some of the funds are slated to send a number of local leaders to the Savory Institute's training centre in Zimbabwe, where local trainers run the courses. An increasing number of villages in Zimbabwe, with similar monsoonal climate and village structures and society but also overgrazed and now desertified landscapes to those in Myanmar, have successfully turned around their landscapes and village societies following such training.

The Myanmar village leaders enthusiastically understand the potential of this new approach in their perplexing and changing world. If the pilot program works (as it has in similar environments and societies in Africa), then the program is set to be rolled out to a further 250 villages, which could involve 35,000 or more people. It seems they have hope again. All going well, this could become their Great Story and part of all our collective Great Work. To me this seems proof that 'what we believe will determine our collective fates': that it really does come down to 'that one square foot of real estate between our ears'.

THE RADICAL HOMEMAKER

CAT GREEN

The year we embarked on the People's Food Plan, 2012, was the year I met Cat Green, who perhaps more than anyone else really breathed life into what was, up to that point, a bold but untested idea. Robert Pekin introduced me to Cat, explaining that he knew this really switched-on person who was keen to get involved in the food movement. Cat and I chatted and straightaway I got a sense of her warmth and gentle humour, but also of her steely commitment to and passion for justice and authentic sustainability.

As the People's Food Plan process took shape, I remember being so excited as yet another public forum was announced and publicised in and around Brisbane and south-east Queensland. Cat and her friends and collaborators organised 12 in total, with over 150 people participating. At the time, I remember telling her how grateful I was to her for helping to bring about a grassroots national conversation about the future of food and farming in Australia, with no external funding and no paid staff. And I remember Cat telling me how grateful she was to me – and AFSA – for creating the space into which she and others could step. And I realised that's one of the many important things that we were doing: allowing others to live their potential, their dreams, to make their special and unique contributions to this growing movement.

The People's Food Plan process resulted in the formation of AFSA's first local chapter – Fair Food Brisbane – again under the initial leadership and impetus of Cat. Then Fair Food Brisbane launched the Garden Dinner Society. And they organised a number of Fair Food Week events in Brisbane in its inaugural year in 2013. You see how these things build: opportunities are created, people get inspired, conversations happen, projects and initiatives are tested. New people get involved, and so it goes. The web expands.

Cat's journey has now taken a special and joyous turn, with the birth of her daughter, Jemima, in early 2014. Her commitment to food system transformation stays strong, and, like all others in this book, the path she has chosen to walk pushes the boundaries of what is possible and what can be imagined in modern Australia.

***

> *Our lives extend beyond our skins, in radical interdependence with the rest of the world.*
>
> Joanna Macy

RENOUNCE: BAD FOOD COMES FROM BAD SOIL

The customs official strode into business class. 'Where is Cat Green? We need to check if we should quarantine the whole plane.'

The other business-class passengers groaned as the customs officials approached me – the scruffy, emaciated early 20-something girl, painfully out of place amid their manicured suits. I was declared safe to enter Australia before being whisked to hospital quarantine for a week. I had the uncommon distinction (uncommon in Brisbane, anyway) of having two tropical illnesses at once and the infectious-disease expert was salivating over my case. After three months of ping-ponging in and out of hospital in Indonesia, my five-year journey of recovery was about to begin.

Renounce. Recover. Rebuild.

These three words are, and forever will be, our journey.

Renounce: I was on a ski trip in New Zealand. My friends mooed at me while we were eating steak. I never ate meat again.

Renounce: While I was living in Syria, Israel invaded Gaza and each nation's news channel told a different story. I never trusted the news to help me understand the world again.

Renounce: They told me hydrolysed margarine was healthy. I never believed conventional nutrition again.

Illness seems to be a common reason for people to become foodies, and my story is no different. While I was ill in Indonesia, I was so sick I couldn't walk. I distinctly remember crawling to the toilet and having to nap on the way there and back (it was two metres from my couch!). My housemate would peel hard-boiled eggs for me before she left in the morning, as this was one of three foods I was allowed to eat. The cleaner one day brought me a porn movie, assuring me it would be appreciated by a bed-ridden Westerner, and I remember spending an inordinate amount of time staring at incense smoke (it looked like dragons). All in all, I was weak. Emaciated. Empty.

When I got the all-clear for typhoid (and was told not to eat shit again) in Australia, I was impelled to explore ways to restore my health. After three months of not walking much, I had lost a lot of muscle. I was in constant pain and deficient in many nutrients. My naturopath later was horrified that I'd never been told to take a probiotic even though I had taken course after course of antibiotics. All I dreamt of was to be well and capable ... and to stay that way!

As I was bumbling my way though these emergent thoughts, I went to the Ideas Festival. As Nick said in his chapter, things often only make sense in hindsight and it's the trivial, random events that propel you in a new direction. For me, this was the Ideas Fest and in particular a session called 'Peak Soil'. Peak soil refers to the loss of our precious topsoil. Over 2 billion hectares of soil have been degraded by human activity over the past 40 years. Topsoil regenerates so slowly (1 millimetre in a century) that it is essentially a non-renewable resource. As *The Ecologist* reported in 2013, 'We are destroying soils worldwide ten times faster than nature can restore them.' But why would that matter ... it's just dirt, isn't it? Well, as I found out, not really.

What struck me most at this talk was the concept: 'Health of soil = health of food = health of you. If the nutrients aren't in the soil, they're not in you.' It's mind-blowingly simple ... and basic ... and revolutionary. I had never thought about where nutrients came from, nor the linkage between soil, food and us. I had only just begun to see food as a potential healing force to replenish my body. After 24 years of just eating food for fuel, I now entered an intimate relationship with food.

Renounce. Recover. Rebuild.

Recover: We started milling our own flour and flaking (rolling) our own oats.

Recover: I started cycling everywhere and my husband maintains our bikes.

Recover: We have three beehives to produce our own honey as a tradeable product.

RECOVER: LEARNING SELF-SUFFICIENCY

When I was a kid, food wasn't something I thought about. My dad made us Vegemite or peanut butter sandwiches for school lunches and either Mum or Dad would cook us dinner. I started a Shapes cracker fad in school simply because I liked them so much, and I think I can also lay claim to the gobstopper craze, but that could also just be my faulty memory. We'd eat out occasionally (McDonald's, of course) and every Saturday we'd wander down to the bakery for a treat. Sunday mornings were pancakes or French toast with maple syrup from my mum's native country and Sunday nights we'd put special music on and have a roast dinner.

It's not exactly food related, but I remember starting a flower garden as a young tacker. The flowers were waving their colourful petals like flags in the breeze until one disastrous day. We were having a pool built and the pool men dumped what felt like hundreds of bags on salt on top of my flower bed ... I can't remember anything more about the incident except that I never gardened after that.

Twenty years later, my husband and I have a vegie patch, backyard chooks, three beehives and a potted fruit orchard. We started cooking as much as we could from scratch – we mill our own flour, flake (roll) our own oats, make yoghurt, cheese and chutneys. I started reading cookbooks for the sheer joy of it – this still surprises me because until I had typhoid and discovered the link between food and health, I can't remember ever being interested in food or cooking. Everything was changing on our quest for more nutrient-dense foods.

We leapt into everything we could to learn. We studied Permaculture and joined a community garden, joined the Brisbane Organic Growers group, joined Transition Towns. We shopped differently, read differently, socialised differently. It's because we were just so curious and willing to learn more about a whole new world that was opening up to us. As John Muir said, 'When we try to pick out anything by itself, we find it hitched to everything else in the universe.' To start gardening is to start learning about seed history and genetically modified organisms, which is to start learning how the body works and about co-evolution of humans, food and bacteria ... and it goes on!

It was during this time of joyous discovery that we had a moment of truth. Not the type that hits you in the face it's so obvious, just the small decisions and realisations of every day. I guess on our path of nutritious health, we could have gone down the path of 'super-foods' and restrictive health diets. I'm not sure we are the type to spend a lot of money on consumables anyway, but we realised that we can't consume our way out of the problems in our world. We need to change how we think of ourselves and our relationship to the planet. None of the superfood fads seem to explore the deeper interconnectivity of the world, or more specifically the link between soil, plants and health. It was a glaring gap, and one that I still struggle with as I try to find the most fresh, recently harvested whole food. It is awkward asking a farmer for a soil profile before I buy their produce (no, I've never done it ... yet), and absolutely impossible in supermarkets. Permaculture, organic gardening and cooking all deal directly with that soil–plant–human nexus and help us have confidence to know exactly what is in the food we are eating.

RECOVER: ACTIVE PARTICIPANTS IN LIFE

When we started exploring wholefoods we didn't plan which skills to learn. We just followed our interests and did what was logical. Case in point – I never set out in life to be a beekeeper. It wasn't something I'd even thought about until a friend asked us to join him one day. I still remember that first time we opened a beehive. The nervousness as we started up the smoker. The anticipation as we prised off the lid. What pierces my memory with clarity is that moment when the lid comes off; the loud hum of 50,000 bees assaults your ears and their incessant movement bamboozles your eyes. Nervousness and anticipation peak as you peer in for your first look. Looking is one thing, but the real thrill is reaching into the mass of bodies to prise out a frame. Absolutely covered in bees, our first act is to shake the bees off the frame. It's bedlam, bees everywhere. They're crawling all over you; they're flying every which way – including buzzing right into your veil; they're walking around on the ground trying to reorient.

During that first hive robbing I was also initiated into the world of beekeepers with my first bee sting since childhood. This initiation might have passed relatively unnoticed had the stinger not become stuck in my ankle for a minute, releasing much of its venom. I ignored the increasing swelling and cycled to work the next day as usual. It was pretty difficult to work that day, not because of the pain but because my swollen leg distracted my colleagues so much. After repeated pleas that I seek medical help, I drew a line on my leg underneath my knee. 'If the swelling reaches this line then I'll go to the doctor.' Ten minutes later I had to go to the doctor where I was prescribed steroids.

The upshot of this initiation was that we now had a unique label for our honey when we sold or bartered it – Big Foot Honey. We accompanied the name with a picture of my engorged leg. It actually looks like it's been photoshopped. Why anyone would take up beekeeping after such a strong reaction I can't tell you, but I was hooked on the thrill and the sweet golden nectar it produced. It's all part of our approach to living, which is to be active participants in our life, not passive bystanders.

You see, life had really started to take a different turn when we began exploring different philosophies such as self-sufficiency, simple living, the concept of 'enough' and, in particular, radical homemaking. Radical homemaking shows how leading a home-centred life enables you to live a life that values family, community, social justice and ecological health. By producing goods yourself and engaging in the sharing economy, you can meet your needs without relying on a paycheque that may (or may not) tie you in with the unethical extractive economy.

Radical homemaking grounds my day-to-day life. It provides the scaffolding onto which I can hitch seemingly unrelated activities – all of which have importance to me. Things like sharing pots of honey with neighbours, crocheting a blanket, writing to my local politician and yarning with a local farmer are all part and parcel of the radical homemaking domain. It is a framework for social change that seamlessly entwines personal change with broader collective change. It makes the personal political, but it doesn't stop there. To be a fulfilled radical homemaker (in my experience and outlined in Shannon Hayes's book of the same name), it's important to go through the rebuilding phase. If you learn the skills but don't engage more broadly with your community, then you're not helping to change the structures and systems that underpin the food system and modern economy.

It was our commitment to these deeper philosophies that opened us up to exploring how to become more actively involved in creating our own lives. We've embraced a production-centred, home-based life that is rich with variety, family, friends, community, nature and meaning. It extends well beyond food to clothing, transport, housekeeping, entertainment, work and our relationship with our community.

Through living this way, we've found that we prefer the variety, simplicity and joy created with a home-centred life of production. We enjoy the integration with the seasons and our community that are at the forefront of radical homemaking. I have found nothing as thrilling in life as doing things you never thought you could ... or simply never thought of at all. Perhaps one of the silliest proud moments I had was when I potato-stamped blue whales on a piece of canvas to use as wrapping paper for a birthday present. It is pretty straightforward but I hadn't really done art since school, and I'm not sure I was that good at it then. So even though it is a basic childhood activity, for me it was groundbreaking. I had made a potato stamp that actually looked like a whale. How cool!

RECOVER: I DEFINE FOR MYSELF WHAT BRINGS MEANING AND JOY TO MY LIFE

Actively building a life that aligns to your values requires skills that are vastly different from those I picked up at school and university. My radical homemaking days have seen me commence my real-life lessons – how to live in kinship with the world (natural and human-made) around me. For my home-based life I've needed to teach myself how to make my own vanilla extract, mead and chutneys. I've learnt a little about chook health, beekeeping and green cleaning. I've sought out teachers for gardening, crocheting and book-binding.

More than just tangible skills, the most important are the intangible skills required to live this way. Radical homemaker Shannon Hayes documents the many necessary skills in her book *Reclaiming Domesticity;* among these are how to build mutually beneficial relationships, self-teach and enjoy what you have. Other intangible skills that I've found important are the art of reflection, being able to let go and the capacity for gentleness (the endless scrubbing of burnt pots is one area of 'tried and failed' that has let me put all these into practice). The fundamental skill that enables us to create our own meaning and live our lives as active participants, however, is the ability to question. A questioning attitude allows us to question social norms and assumptions about the way things are done. It allows us to determine for ourselves what is our 'work that matters', rather than having our pathway defined by someone else.

Being able to question the 'givens' in our society has helped me be open-minded about what actually brings me joy and what is my 'work that matters'. When embedded in a society that equates 'needs' with 'consuming', it is very difficult to walk a line that values making and doing over consuming. I've found the passive act of consuming has brought me very little joy, whereas the hands-on skills, as well as the intangible skills, have brought me immense fulfilment and joy. My 'work that matters' comes from being actively involved in life, not sitting on the bleachers paying for someone else to do it.

RENOUNCE: PART OF THE WHOLE

So if our first moment of truth was realising we wanted to live actively through making, creating and connecting, then our second realisation was a bit different. We were ensconced in this beautiful community of people and we were making, bartering and supporting businesses we believed in. We were learning and sharing and attempting to live with 'enough'. Life was full of joy and meaning, and yet it was often a struggle when our lives collided with mainstream culture. This was particularly so in relation to sourcing fresh, nutritious local food.

My primary food values are aligned with Food Sovereignty. The food has to support people, animals, producers and the environment ahead of corporate profit. And for me, a fifth imperative was for the food to be nutrient-dense. Despite living in the inner city with an abundance of alternatives around me, I realised that I am part of the whole. If it is hard for me, it is because it is hard for everyone.

We all share the land, water, air and soil as our commons (although that isn't realised in our privatised marketplace). And so what affects one affects all. My struggle to source ethical food is endemic of the system in which it is a struggle to *produce* ethical food.

I see how hard it is for producers – they battle the weather, pests and volatile commodity markets. They have to endure a monopolistic distribution and retail industry in Australia, which gives more power to those further up the supply chain and producers little option but to sell their produce on terms dictated to them. I see how many farmers require off-farm income to keep their operations afloat and how many are encouraging their children to leave the land.

I see how hard it is for ethical retailers – when conventional food is so cheap at the supermarkets, it can be very difficult for ethical retailers to pay the farmer well, cover their operational costs, keep their food affordable and to make a living wage. I see businesses struggle to balance these while competing for market share from Coles and Woolworths who have been hauled, time and time again, before the Australian Competition and Consumer Commission on allegations of misconduct.

I see how hard it is for eaters – sourcing ethical food is challenging. And it takes so much time! Because I am passionate about this I make time; however, I understand others don't have the time or interest to do this. They either don't care, or want to be ethical but don't want to put any effort into it. And in most suburbs in Brisbane that's difficult if not impossible, because often there's only one option – Coles or Woolworths. In West End, where I live, there is an abundance of local businesses and food co-ops that are within a short walk and that sell organic, local, ethical produce. On the odd occasion I go to Coles, I am bewildered, befuddled and bamboozled by the sheer amount of options that I don't consider to be real food. In fact, there's very little in Coles that meets my five food requirements. So where would that leave me if I left my inner-city enclave? I would have very few options readily available.

RENOUNCE: I COULDN'T BE ETHICAL IN AN UNETHICAL SYSTEM

I realised that I couldn't be ethical in an unethical system and I wasn't okay with that. Here is perhaps the moment where life could have gone differently. I could have felt so powerless, despairing and disenfranchised that I cuddled up to a ready-cooked, frozen meal. I could have felt it wasn't worth the sacrifice of time and effort that would be involved to change the system. I could have become busy with a career, my social life or any number of things. But I didn't.

As an old Chinese proverb says, 'to know but not do, is not to know'. I knew, so I decided to work for something better. You see, it wasn't just about me anymore. What had begun as a quest for health had expanded. I now cared about the farmers; the soil and its teeming multitudes of life; the trees; the birds and bees; the chooks who can't flap their wings; the small businesses who are preyed upon by supermarkets' predatory business practices; and my fellow residents, who daily lack fresh food and a connection to the land where fresh food is produced.

Renounce. Recover. Rebuild.

Rebuild: I joined the Australian Food Sovereignty Alliance to build and advocate for a Fair Food system for Australia.

Rebuild: I sought out others to share knowledge, skills and friendship.

Rebuild: Our friends buy us ethical toilet paper when we can't make it to our bulk buyers club.

So how do we support the whole system to which we are ultimately and intimately linked?

I met a woman at a Palestinian refugee camp between the borders of Iraq and Syria where I was documenting refugees' stories. She had been university educated and loved relaxing in a warm bath. She had once had a career, hobbies and a family she loved. Then her house got bombed and she fled Iraq. What little she had accumulated at the refugee camp had just been damaged when the camp's sewerage system backed up in the harsh winter and flooded her tent.

She wanted to live in peace and safety. She wanted a chance to use her skills and contribute. At the end of the day, all she wanted was a life as an active participant.

It was through her story, as well as many other stories and experiences I've had, that I realised complex problems aren't solved by one person. This applies to war and refugees, and it applies to our corrupt food system. To tackle power we need to accumulate power.

REBUILD: ACCUMULATING POWER

Accumulating power is not an armchair exercise. I know because I've tried it. No matter how many self-righteous rants I had over drinks at my local, the world wasn't changing just because I was angry at it. I think tangible, systemic change is created when we accumulate power through actively working together.

I know on the face of it, it might seem we have more pressing problems than worrying about food. An 80 per cent supermarket duopoly in Australia is nothing compared with the changing climate, or the slave labour trade ... but isn't it? That's the curious thing about the food system – it's not actually just about food. It's about the environment. It's about the rights of animals and other species. It's about justice. It's about health and wellness. You see, food isn't just a commodity, or a calorie needed for refuelling. Food is nutrition. Food is culture. Food is tradition. Food connects us with the earth that sustains us. Food connects us to our bodies. Food connects us to each other.

For this reason, it is an issue that unites us. We might all come to Fair Food activism for different reasons, but we are all fighting the same Goliath.

If there was ever to be a struggle that mattered to all Australians, it would be the struggle over food, and this means that Fair Food activism provides a unique opportunity to pool our collective power. We can build a united and coordinated resistance that is actually so powerful that it creates a new system simply by the sheer number of people involved ... or so is the hope anyway.

RENOUNCE, RECOVER AND REBUILD: FLOUR POWER

Before I embarked on this wholefoods journey, flour was just flour to me. The only difference was whether it was plain or self-raising, and I couldn't have told you what actually made it rise. I didn't know where, how or when wheat became flour and couldn't have cared how long it sat on the supermarket shelf. I had certainly never heard of sourdough bread or the seemingly endless types of flour such as coconut flour, almond flour, rice flour, etc.

These days I buy biodynamic spelt berries in 20-kilogram bags from a local farmer distributor. We then mill our own flour as we need it. We try to not mill it more than a week in advance as this causes the flour to lose most of its nutrients and go rancid. I am learning more and more about traditional preparation methods as well as wheat's role as a rotational crop and local, place-based varieties.

I started doing all this for personal health reasons. Store-bought flour is rancid and devoid of nutrients. Not only does it not nourish my body but it robs it of nutrients and energy absorbed from other food! While fresh stone-ground flour or white flour sourdough makes it less of a devil (the 'anti-crust'), it is still a far cry from the body-nourishing food that has made it a staple food.

Despite grains having a bad reputation in some circles, for me, flour is a potent symbol of the type of personal and systemic transformation that's required. I have gone from being a disinterested bystander to having an intimate relationship with flour – not just as a source of food, but with the history, tradition and culture that surrounds growing and preparing it. I believe if we all approach food open-heartedly and are prepared to embrace this totality, then our food system will naturally evolve towards supply chains that are rooted in people and place. This is when producers and eaters will work collectively to sustain and enrich our food and agricultural system.

REBUILD: LOCAL DECISION-MAKING

Experience has taught me to trust in the basic goodness of people, and I think we all essentially want the same things. We all want clean air, clean water, clean land and good food. We want to be happy and healthy and we want that for our kids and grandkids. So it seems to me that people like us aren't involved in decision-making around food at the moment, otherwise we wouldn't have policies such as spraying dangerous chemicals, advertising junk food to kids and using prime arable land for fracking.

For me personally, one of my guiding principles in life is that people should be able to control decisions that affect their lives. Said another way, we all need to be actively involved in our own lives and the life of our community. Collective decision-making is at the heart of Food Sovereignty, and the reason I jumped on board with the Food Sovereignty movement in Australia is because I found that it spoke to something more than the sum of its parts. I had been involved in many local food groups, all of which were fantastic and important initiatives, but, for me, something was still missing. Each of the individual groups or projects was tackling an individual problem, but none were addressing the systemic issues that produced all these problems. That systemic issue is that the food system is based on values that put corporate profit ahead of humans and places. I think the reason that this has occurred on such a deep and systemic level is because people aren't actually able to control decisions about our food and agricultural systems.

RECOVER: I AM MORE THAN JUST A CONSUMER

I've had lots of experiences of 'being participated' instead of being an active participant. Focus groups where your ideas are taken but you never know if it made a difference. Online petitions, I think, are similar – you share your concerns but does it actually impact decision-making? And another common one is using my 'consumer power'.

One of the books I read on my food learnathon was Michael Pollan's *Food Rules.* It's a terrific mingling of old wives' tales, community wisdom and modern science presented in catchy layperson language such as:

* don't eat food with ingredients that you can't pronounce
* don't eat anything your great-grandma wouldn't recognise as food
* don't eat food that won't eventually rot.

With these and other principles in mind, I changed what I bought and tried not to shop at Coles and Woolworths. But to tell the truth, I'm not sure they've noticed. In part, I think we're all being sold a fairy tale – that of the mythically powerful 'consumer'. Realistically, I think this whole idea of consumer power and 'vote with your dollar' is very limited. It restricts us to a singular role – that of being a consumer.

On a practical level, consumer action is incredibly complex and complicated on a day-to-day basis. In my studies, I had seen how schemes like fair trade and organics started off with good intentions, and worked tolerably well; however, there were significant drawbacks. Just one example of this is my favourite Green & Black's fair trade, organic dark chocolate. I thought it was an ethical purchase (and in some ways it still might be the best, given its ethical purchasing standards) until just recently when I learnt it is in fact owned by one of the world's largest multinational food corporations, Mondelez (formerly Kraft). If we hadn't googled it, we'd never have known from the packaging. The current lack of transparency and understanding of how companies operate make consumer power really difficult to enact.

At a more philosophical level, the idea that my most potent weapon for food system change is as a consumer offends me. Remember, I am a radical homemaker, living a home-centred life. In radical homemaking, the more I can produce for myself and my community, the more I am able to contribute to an ethical, life-serving economy. However, under the consumer power model, if I'm not consuming then none of my efforts and energy are valued. It also marginalises those who don't have enough money or time or knowledge to shop ethically. The vote-with-your-dollar concept further entrenches the neoliberal capitalist system, which sees us only as bags of money on legs, not as the heart-beating humans we are.

Having said all that, I do strongly believe that we need to shop wisely and support ethical businesses. How are we going to create new solutions if we don't support people who are currently trying to operate businesses on different values? Even as we use our money to support local, ethical businesses, we need to know the story doesn't end there. We are more than just consumers, and consumer power alone will not change the system. It might make it a bit better, sometimes, for some people, but it doesn't actually create a system that's built on different values. In order to create a Fair Food system, we all need to have the opportunity to be actively engaged in democratic decision-making.

REBUILD: PEOPLE'S FOOD PLAN

In the case of food system transformation, what's required is genuine participation. Patricia Murphy and James Cunningham have defined this as 'a process whereby the people of a community, regardless of income or position, join meaningfully in making social, political and economic decisions related to the general affairs of the community'. I wanted to authentically and meaningfully participate in building a better food system. I was itching for something to do, and then the opportunity came. Enter the People's Food Plan. This is when I really got involved with the Fair Food movement.

The People's Food Plan was a proactive, solutions-focused, democratic undertaking in which everyday Australians got to have a say in the kind of food and agriculture system they wanted. It was Australia's first crowdsourced public policy document. Not only was there something I could get in and do, but that something was to invite others to become active decision-makers too!

I got involved with the Fair Food movement when AFSA was about to start the public consultations for the People's Food Plan. I threw myself into the organising in Brisbane and south-east Queensland. Working with other community organisers, we arranged

12 forums over three months that were attended by over 150 people. These forums were a chance for people to engage and debate the kind of food system they wanted. Using the People's Food Plan draft paper, people were able to review problems and options for future direction. There were as many different ideas as there were people, but at the forums I was involved in, all were in agreement with the fundamental tenets of Food Sovereignty – that a healthy food system looks after humans, animals, producers and the environment ahead of corporate profit.

RECOVER: TIME FOR THINGS THAT MATTER

I am fortunate to be in the position to be able to pursue important community initiatives. With our radical homemaking and beliefs in 'enough', we are able to make room for 'work that matters'. To me, it doesn't matter if this 'work that matters' is paid or unpaid. There's not really such a distinction in my life anymore. As long as we have enough to get by, I don't need a salary to know what I do is valuable. What's more important to me is that I believe my work is valuable. To be able to say that my hours were put towards creating something bigger and better for everyone is to be able to say I've had a good day.

I think that is the crux of my journey and how my Fair Food activism fits seamlessly into my life. It is not an extra; it is just part of it. Something as necessary as mucking out the chooks or soaking the mung beans for sprouting. Not to say that it always takes priority either, everything has its season, but I know it will always be a part of my life. 'Work that matters' doesn't just disappear when the funding dries up; you can't be made redundant.

REBUILD: CONNECTEDNESS

After the People's Food Plan, a group of us wanted to continue working on Food Sovereignty in Brisbane. We set up Fair Food Brisbane as a sub-branch of AFSA in 2013. The founding and early members of Fair Food Brisbane are Ian Kirkland, Hope Johnson, Rachel Ryan and Aislinn Grech, with Louise Abson, Arlanda Rayne and Jemma Walden joining us as well. Working with dozens more passionate people over the past two years, we've been able to host numerous events. Fair Food Brisbane is made up of terrific people, working well together to get things done.

The defining characteristics of the modern food system are atomisation and disconnection. Uncoupled and untethered, we can never see food for what it really is – the culmination of thousands of years of tradition, history, culture and connection to a place. A Fair Food system is a connected system and this was the early mission of Fair Food Brisbane.

Through the Garden Dinner Society, Fair Food Week events, ethical cafe social media competition and the *Just Food* documentary launch, connection is at the forefront of what we do.

We can see this through the lens of the Garden Dinner Society. It focused on connectedness between farmers and eaters, between eaters and seasons, between eaters and other eaters. We wanted to host events that were ethical, beautiful and would inspire the magic of connection. Motivated by Dîner en Blanc's impressive waiting list in Brisbane, we started with the idea of a 1000-person dinner. Early on, we realised we had no idea what we were doing and should start small. Success at a small scale would lead to more success, but a public event of 1000 people was a disaster – or at least stress-induced ulcers – waiting to happen!

To date we have hosted three dinners in community gardens. The Jane Street, Jeay Street and Inspiration Community gardens were poignant settings in which to focus on food's connectedness to the land. Each dinner was preceded by a garden tour and a discussion on seasonality and the place of community gardens in building a Fair Food system. All our chefs – George Clegg of Malt, Bryant Wells of Tukka and Shannon Kellam of the Brisbane Club (as well as being Australia's representative to the French food Olympics, Bocuse d'Or, in 2015) – were terrific ambassadors for seasonal, local food. We included as much garden produce on the menu as was possible, with the remainder of the food being sourced ethically and locally (with George even providing pumpkins from his mum's garden!).

Shannon Kellam gave a memorable talk about the trends in the chef world. He said that in the 1980s the fashion was to really work the food, so the quality and flavour of the produce was less important because it was hidden and lost. Nowadays, the trend is for simple, pared-back meals that feature the produce. He said he spends two to three hours per day chatting with producers, trying to source the best quality ingredients. Agricultural practices, timing of harvest and whether it's chemical or organically produced all have an impact on quality and flavour.

At the close of 2014, Fair Food Brisbane was at a cross-roads – there is a whole big world of food system change out there, and it was hard to know where to put our energy to have the most impact. As an entirely volunteer-run organisation, it's also difficult to sustain the energy and resources needed for systemic change within the complex web that is life. Family, work, community and activism are all part and parcel of all our lives and each aspect will be the focus at different times. This isn't unique for Fair Food Brisbane, nor the Fair Food movement. It is, however, an interesting conundrum as we work together for systemic change and collective decision-making.

By 2015, we had started focusing on research. Again, realising that the food system is a giant beast to change, we were besieged by ideas and innovations from other places. These were incredible initiatives, but would they work in Brisbane? What did the food system actually even look like in Brisbane? And so we've gone back to basics in an effort to really know and understand our neck of the woods. Partnering with the wonderful farmer-focused organisation Reciprocity, we are starting our research journey by trying to understand the number and nature of farms and farmers in our local area. This research will be available to anyone who is interested, and from Fair Food Brisbane's point of view, it will help us see where best we can put our efforts for systemic transformation.

REBUILD: A SOCIAL MOVEMENT OF BEING

If I started my chapter with my particular personal story, then I end it musing on the nature of what it means to be human. We humans are a species like all others. Like all species, we are born, live and die. We are nourished by the sun, air, water and soil. We are reliant on the earth for our daily sustenance. Because of our intimate relationship with the natural world, what happens to the sun, air, water and soil is what happens to us. To know this is to know that we are part of nature and it is a part of us.

Food is the ultimate embodiment of this relationship, and so to call for food systems change is, in essence, to call for a change in how we see ourselves as humans. I think food should become the 'gateway drug' for bioregionalism and our recovered kinship with the natural world. Food is not only the tangible product of a place, but the metaphysical language, the history, culture and tradition of that place.

I think we as humans, and definitely as food activists, need a different worldview, different values, different skills and ways of being. I don't believe it is enough to want something different. It is not enough to speak for something different. We are the ones that need to be different. For me personally, this has come from using radical homemaking as a framework to help me see differently and learn to be differently. I am using chutneys, gardening and neighbourhood potlucks as a source of connection and meaning. Other people will find their own meanings and their own journeys.

As long as we are all on a journey together, exploring what it means to be human and to live in kinship with our earth, then we will see change. So rather than just calling for a fairer vision of an interconnected food system, I am calling for a fairer vision of an interconnected world. Ultimately, I am calling for a social movement of being.

CONCLUSION

NICK ROSE

The stories in this book are stories of commitment and passion, of individuals who have chosen to dedicate their lives to something bigger than themselves.

These are stories of courage, of people who have confronted feelings of inadequacy and anxiety, and faced their own fears and demons.

These are stories of people who are not afraid to take risks, to move beyond and outside of their comfort zones, and see where such daring takes them.

These are stories of reflection and self-examination, of people who have considered deeply what it means to be 'human' at this point in our evolutionary journey, what their particular skills and capacities are, and what life choices they can make in order to work for the betterment of their families and communities.

These are stories of change, of people who have made major shifts at key moments in their lives, shifts that have set each of them on a course of personal transformation. Shifts such as the momentous decision to become a farmer, or a food forest gardener, or a radical homemaker; or the decision to farm in a different way from one's father, or grandfather, or neighbours; or to leave their country, and embark on further study leading to a totally new direction in life. To embrace a new identity as an 'activist', and dedicate a major portion of their time and energy to raising awareness about critical issues of our times, such as the ethics of raising animals for food in confined spaces, or the relentless and rapid expansion of genetically modified technologies, seemingly without pausing to consider the consequences or the risks. Or to embrace a new identity as a 'social entrepreneur', or even a 'foodpreneur', focused on creating new ways to facilitate the scaling up of a local and Fair Food system, such as the model of Food Connect, or the open source distribution software of the Open Food Network.

You have read the stories of 11 individuals, each of whom I would describe as a Fair Food pioneer: someone who realises that the challenge of our times is to live in ways that are consistent with Thomas Berry's admonition to all humanity, 'to be present to the planet in a mutually beneficial manner'. The really good news is that while these individuals are all remarkable in terms of their vision and achievements – and in that respect can and should be described as leaders – they are far from alone. On the contrary, they are representative of a swelling multitude of people around this country, and around the world, who are living and acting in ways that expand the boundaries of what is possible, what is feasible, what is acceptable.

I know, because I have met so many people I would describe in these terms.

All my experiences in Australia and internationally over the past eight years – from the humble beginnings of a local food network in a small regional town in northern New South Wales, to the establishment of a growing national organisation with direct links to a global movement of 300 million people – reminds me of the conversation I had with Nettie Wiebe, the first female president of the Canadian National Farmers Union and one of the original founders of La Via Campesina, during the course of my PhD research. We were talking about whether there were grounds for 'optimism' or 'pessimism' with regards to food system – and social – transformation, and this is what she had to say:

> I find myself less than optimistic when I note the current trend lines – but I am always hopeful ... Optimism would have to be based on some realistic assessment of the trend lines, because optimism is based on looking forward on the basis of what you already know about, or can reasonably expect. Whereas hope is the imagined possibility of something better.
>
> So when we look at the trend lines, they don't look very happy. We continue to lose biodiversity. We are continuing to have major and minor, but a multitude, of changes going on, very few of which look like they're enriching and enhancing

> the life-systems, most of which look like they're destroying and undermining the life-systems on earth. Whether it's old-growth forests, or salmon runs, or soil degradation, or birds of prey, or buffalo herds – you pick your favourite item to see whether it's flourishing or not. The database on all of this, including on the numbers of small farms and farmers, doesn't give you much room for optimism.

Then we talked about 'hope':

> On the other hand, every time you put a seed in the ground, and it comes up, and it flourishes, and you get far more from it than you had expected, that's a kind of minor miracle. We live in a world that has those miracles around us, and that for me is an opening to imagine that we could do [better] ... [And] it's not just in the biological world; every time I sit in a room full of people whom I have not known before, and whom I then discuss with, and hear, and sing with, and find common ground with, and I feel the power of our togetherness and the possibility that that offers, I say to myself, 'There's so much hope.' There's such a huge prospect of imagining and working towards, and finding, sometimes by miracle, a better world.
>
> I can be pessimistic, but I can't be despairing, I can't be hopeless.

I know exactly what Nettie Wiebe is referring to here, because I too have often sat in rooms of 'strangers', and heard stories of passion and action and change. On 11 March 2015, I felt shivers of delight and possibility as I listened to Scott Callaghan, teacher at Mount Austin High School in a low-income district of Wagga Wagga, New South Wales, talk about how the decision to convert the five-acre school agricultural plot from running a few sheep to a horticulture plot was transforming the lives of those it touched. This included the students, 95 per cent of whom were now taking home fresh produce, whereas only 5 per cent previously were taking home some of the sheep's wool. It included the clients of emergency food providers in the city, who, as a result of the school's participation in the council-facilitated 'Grow and Give' program, were now receiving fresh local produce. And – perhaps most exciting of all – it included the shoppers at the local IGA. In the summer of 2015, the plot was producing so much corn that it approached the IGA to see if it wanted to buy some. The IGA said yes, and is now selling over 700 ears of corn a week, whereas before it sold an average of 20.

Why? Because the local residents love being able to eat fresh local produce, and most especially when it's being grown by their own sons and daughters in the local high school. Those Wagga residents respond to the connection and intimacy of that produce. My intuition tells me that's what *all of us* want – that kind of connection and intimacy.

When Scott related that story to the local audience who'd come to watch the screening of *Fair Food,* he received very warm applause.

As I have heard these stories – and there are so many more – I realise that more and more of us share a vision of a better world, a fairer world, and that we will do what we can to bring it about. Bringing that better world about is a process, and we don't know what progress we can make in our lifetimes. But this work, of building and sustaining a social movement focused on large-scale change, which begins from the smallest actions such as growing your own herbs and greens, is as much about the journey as it is about the destination.

There are no guarantees of 'success'. Working together can be and often is difficult and generates conflict and tensions, especially in highly individualistic cultures like Australia, where the focus of life is very much on 'me' and 'mine' and 'I'. Yet, as Nettie Wiebe says, working to build a social movement is richly rewarding and invigorating. It's a life of connection, and connectedness. It's a life that nurtures a deep sense of inner peace, and joy, because it is a life of meaning. It's the realisation, at a spiritual level, that if we connect to something larger than ourselves, we have within us the power to make a real contribution. Each of us has the power to make a difference.

And all of us want to do that – don't we?

ACKNOWLEDGEMENTS AND DEDICATIONS

I firmly believe that almost all human endeavour is collective and collaborative in nature. As Albert Einstein, truly one of our moral and intellectual giants, wrote over 80 years ago: 'A hundred times a day I remind myself that my inner and outer lives are based on the labours of other people, living and dead, and that I must exert myself in order to give in the same measure as I have received and am still receiving.'

This book has been years in the writing, and has involved hundreds of people across continents and hemispheres. I feel enormously grateful to have lived the life that I have, and that I have been able to travel so far and meet so many inspiring and wonderful people. My 'riches' are in my relationships, and those relationships have allowed me to be the person I am today and become the person I will be tomorrow.

I am grateful to every one of the authors, each of whom has made time available in their very full lives to share their experiences in the pages of this book. I am grateful to David Pocock and Guy Grossi, for contributing the forewords.

I am grateful to Charlie Massy for introducing me to Alexandra Payne, and to Alex, and the editorial staff at UQP, for believing in the project of this book, and the cause and ideals it represents. I am grateful to Nikki Lusk for her masterful and supportive editorial work in crafting all these diverse chapters into a coherent whole.

I am grateful to every person who buys and reads this book. My wish is that within its pages there are words and stories that speak and resonate with you.

Most of all, I am grateful to my partner, Julie Tucker, who, after six years of supporting me through the researching and writing of a PhD, and no small amount of unpaid community activism, embraced this project with great enthusiasm. Her reserves of patience and understanding know no bounds.

I dedicate this book to my mum, who died during its writing, at the very Australian age – at least for a lover of cricket – of 87. Hers was a life of decades of service as a nurse, and boundless energy in raising a family of six. After a very long time away, and years of estrangement, I am happy and proud to call myself her son.

I dedicate this book to the memory of the hundreds of thousands of Guatemalans who perished during the 36 years of the brutal internal armed conflict that country endured from 1960 to 1996. Most of them died hoping and dreaming that one day their lands would know peace, and freedom from poverty and oppression.

My hope is that I can make a small contribution to bringing that day closer, not just for Guatemala, but for all of us. And so I also dedicate this book to my two sons, Camilo and Jude, and to all your generation. May the Australia you inherit be a beacon of peace, equality and fairness for all.

ABOUT THE CONTRIBUTORS

Robert Pekin's agricultural journey transitioned from fourth-generation organic dairy farming to pioneering the community-supported agriculture movement in Australia. In 2004, he founded Food Connect Brisbane, a dynamic multi-farmer local food hub. Later, the Food Connect Foundation was launched to support the Fair Food movement, and he is a member of many inspiring Food Sovereignty organisations. He loves his blended family of four children, and is passionate about micro food enterprises that produce beer, butter and artisan bread.

Emma-Kate Rose has an extensive background in applied research and project management. She joined Food Connect as General Happiness Manager in 2011, coordinating a significant internal restructure. In May 2013, Emma-Kate became a Fellow of the Australian School for Social Entrepreneurs. She is passionate about working with the community to bring greater awareness to the issue of Food Sovereignty. She loves her city cowboy, Rob Pekin, who helps her with the kid wrangling.

Michael Croft is an agroecological farmer and was the inaugural and immediate past president of AFSA. He is an Australasian civil society delegate to the UN's Committee on Food Security, an Oceania delegate to the International Planning Committee for Food Sovereignty, an international advisory board member to the More and Better Network (Norway), an advisory member of Food Tank USA, a Slow Food convivium leader and Terra Madre delegate to Italy, a board member of the Australian Slow Food Ark of Taste, a guest lecturer at the Fenner School at the Australian National University and a working party member for the Institute of Ecological Agriculture Australia Association. He has a graduate diploma in Rural Leadership from James Cook University and is a Fellow of the Australian Rural Leadership Foundation.

Angelo Eliades is a presenter, trainer and writer in the areas of sustainable gardening and Permaculture, and a passionate forest gardening advocate and designer. His award-winning demonstration food forest garden in Melbourne is regularly open to the public. When he is not consulting, teaching, working in the garden nursery industry or enjoying time in his own garden, he supports community gardening projects and runs the educational website *Deep Green Permaculture.*

Kirsten Larsen has been working on and in food systems for over eight years. Her work at the Victorian Eco-Innovation Lab at the University of Melbourne has focused on innovations to develop more sustainable and resilient food systems. Kirsten's policy background includes work for state and local governments on climate change and food policy. Kirsten is a Director of Eaterprises Australia, a social enterprise focused on transforming the way food is produced, distributed and exchanged. She is a co-founder of the South East

Food Hub, the Australian Food Hubs Network and the Open Food Network – an open source platform empowering interconnected communities of food producers, distributors and buyers to co-create new food systems.

Fran Murrell first learnt about GM crops when her first child was aged one. The promises of reduced pesticide use made GM crops sound attractive, but the more she investigated the more concerned she became. In 2007, she co-founded MADGE Australia Inc., a group of mothers and others trying to sort through the spin on GM food. She helps MADGE to research the food system to allow people to choose food that is good for those who eat it, grow it, produce and sell it, and for the land and environment. She has spoken on this topic nationally and internationally.

Carol Richards is a Senior Research Fellow based in the School of Management, Business School, Queensland University of Technology. She specialises in food, agriculture and resource sociology, with an emphasis on food value-chain governance, corporate land acquisition, and climate change and new social movements. She was vice-president of the Australian Food Sovereignty Alliance from 2013 to 2014 and convenor of the Australasian Agri-Food Research Network from 2010. She has published on the topics of sustainable natural resource management, food security and supermarket power and is currently researching the fossil-fuel global divestment movement.

Tammi Jonas is an ethicurean farmer, mindful meatsmith, and agrarian intellectual at Jonai Farms & Meatsmiths, where along with her husband, Stuart, and three politically engaged children she raises rare breed Large Black pigs and cattle on volcanic paddocks near Daylesford, Victoria. In 2013, Jonai Farms was the first farm in Australia to wholly crowdfund major agricultural infrastructure, which is now a thriving retail butcher's shop and salumeria on the farm. Tammi has been writing about food ethics and politics since 2006 on her blog *Tammi Jonas: Food Ethics,* and works publicly for a democratic, ethical, and regenerative food system in her role as president of the Australian Food Sovereignty Alliance. She can be found on Twitter and Instagram as @tammois, and for the farm as@jonaifarms.

Charles Massy (OAM) recently completed a PhD at the Australian National University, examining transformative change in regenerative agriculture (where he now lectures in Sustainable Rural Systems). Charles has managed the 1820-hectare family sheep and cattle farm for 40 years, and still consults widely in the fields of Merino breeding and transformative change in agriculture. He has chaired and served as a director with a number of businesses, research organisations, and international manufacturing and marketing organisations. Charles has published widely in a range of fields, including three books, and is currently writing a book on regenerative agriculture, to be published by UQP in 2016.

Cat Green is a radical homemaker, dwelling in her home town of Brisbane. Cat is an active participant in her home and community and thrives on trying to produce more of what she needs. By turning her home into a centre of love and production, Cat is living her values of looking after people and planet. She has previously travelled, worked and studied in the social change field.

FURTHER READING

FOOD SOVEREIGNTY/FAIR FOOD

For the history of AFSA, including its founding open letter to Tony Burke in August 2010 and the People's Food Plan: www.australianfoodsovereigntyalliance.org

For more detail on the global Food Sovereignty movement, see La Via Campesina's 2013 *Jakarta Call:*www.viacampesina.org/en/index.php/our-conferences-mainmenu-28/6-jakarta-2013/resolutions-and-declarations/1428-the-jakarta-call

For more detail on agroecology, see the Nyeleni Declaration of the International Forum for Agroecology, released on 27 February 2015: www.foodsovereignty.org/forum-agroecology-nyeleni-2015; and Olivier De Schutter, UN Special Rapporteur on the right to food, *Agroecology and the Right to Food:*www.srfood.org/en/report-agroecology-and-the-right-to-food

Committee on World Food Security: www.fao.org/cfs/cfs-home/it

CSM (international food security and nutrition civil society mechanism): www.csm4cfs.org

Food Connect manifesto: www.foodconnect.com.au

Open Food Foundation: www.openfoodfoundation.org

Open Food Network: openfoodnetwork.org

Reciprocity: www.reciprocity.org

Slow Money: www.slowmoney.org

FOOD AND SOCIAL TRANSFORMATION

Thomas Berry, *The Great Work: Our Way into the Future,* Broadway Books, New York, 2000.

Wendell Berry, *Bringing It to the Table: On Farming and Food,* Counterpoint, Berkeley, CA, 2009.

Doyle Canning and Patrick Reinsborough, *Reimagining Change: How to Use Story-Based Strategy to Win Campaigns, Build Movements and Change the World,* PM Press, Oakland, CA, 2010.

Douglas E Christie, *The Blue Sapphire of the Mind: Notes for a Contemplative Ecology,* Oxford University Press, New York, 2012.

Shannon Hayes, *Radical Homemakers: Reclaiming Domesticity from a Consumer Culture,* Left to Write Press, 2010.

David Korten, *The Great Turning,* Kumarian Press, West Hartford, CT, 2007.

Bruce H Lipton and Steve Bhaerman, *Spontaneous Evolution: Our Positive Future and a Way to Get from Here to There,* Hay House, Carlsbad, 2010.

Charles Massy, *Transforming the Earth: A Study in the Change of Agricultural Mindscapes,* ANU PhD thesis, 2013.

Donella Meadows, *Leverage Points: Places to Intervene in a System,* The Sustainability Institute, Hartland, VT, 1999.

Patricia Murphy and James Cunningham, *Organising for Community Controlled Development: Renewing Civil Society,* SAGE Publications, London, 2003.

Michael Pollan, *Food Rules: An Eater's Manual,* Penguin Books, London, 2009.

EF Schumacher, *Small Is Beautiful: A Study of Economics as if People Mattered,* Vintage, London, 1993.

Richard Wilkinson and Kate Pickett, *The Spirit Level: Why Equality Is Better for Everyone,* Penguin Books, London, 2010.

GARDENING AND FARMING

Dan Barber, *The Third Plate: Field Notes on the Future of Food,* Brown Book Group Limited, Little, 2014.

Charles Benbrook et al., 'New Evidence Confirms the Nutritional Superiority of Plant-Based Organic Foods', The Organic Center, March 2008: www.organic-center.org/reportfiles/Nutrient_Content_SSR_Executive_Summary_2008.pdf

Angelo Eliades, *Deep Green Permaculture: The Sustainable Organic Gardening Guide for Self-Sufficient People:* deepgreenpermaculture.com

Bill Mollison and David Holmgren, *Permaculture One: A Perennial Agriculture for Human Settlements,* International Tree Crops Institute USA Inc., Davis, CA, 1981.

Joel Salatin, *You Can Farm: The Entrepreneur's Guide to Start and Succeed in a Farming Enterprise,* Polyface, VA, 1998.

Allan Savory, *Holistic Resource Management: A Model for a Healthy Planet,* Island Press, Washington, DC, 1988.

EXPOSÉS OF INDUSTRIAL FOOD AND AGRICULTURE

For more details on genetic engineering and modification, see MADGE Australia Inc: www.madge.org.au

Tim Flannery, *Here on Earth: An Argument for Hope,* Text Publishing, Melbourne, 2010.

Felicity Lawrence, *Not on the Label: What Really Goes into the Food on Your Plate,* Penguin Books, London, 2004.

Kristen Lyons, Carol Richards and Peter Westoby, *The Darker Side of Green: Plantation Forestry and Carbon*

Violence in Uganda, The Oakland Institute, November 2014: www.oaklandinstitute.org/sites/oaklandinstitute.org/files/Report_DarkerSideofGreen_hirez.pdf.

Charles Massy, *Breaking the Sheep's Back: The Shocking True Story of the Decline and Fall of the Australian Wool Industry,* UQP, St Lucia, Qld, 2011.

Oxfam Australia, *Banking on Shaky Ground: Australia's Big Four Banks and Land Grabs,* April 2014: www.oxfam.org.au/wp-content/uploads/site-media/pdf/2014-47%20australia%27s%20big%204%20banks%20and%20land%20grabs_fa_web.pdf.

Lynn White, 'The Historical Roots of our Ecologic Crisis', *Science,* vol.155, 1967, p.1203.

Donald Worcester, *The Wealth of Nature: Environmental History and the Ecological Imagination,* Oxford University Press, Oxford, 1993.

BACK COVER MATERIAL

Australia's food system is more than just broken: it's killing us. Now is the time to act, to make a difference – to change the world.

The groundbreaking *Fair Food* tells the new story of food: how food and farming in Australia are dramatically transforming at the grassroots level towards reconnection, towards healing – of the land, of each other. It offers a compelling and coherent vision of how our future can be so much better than our present and our past, and how each of us can make a difference.

Told through the experiences of several of the leading figures in Australia's Fair Food movement, this book tells stories of personal change, courage, innovation and food activism, from local food hubs and backyard food forests, to the GE-free movement, urban farming, radical homemaking and regenerative agriculture.

In a time of bullying corporations, supermarket monopolies and environmental degradation, *Fair Food* offers compelling and inspiring stories of personal transformation from everyday people, showing us that we, too, can be powerful agents of change in this time of need.

Edited by Fair Food pioneer Nick Rose, and with forewords by David Pocock and Guy Grossi, the book includes contributions from **Michael Croft • Angelo Eliades • Cat Green • Tammi Jonas • Kirsten Larsen • Charles Massy • Fran Murrell • Robert Pekin • Carol Richards • Emma-Kate Rose**

Printed in Dunstable, United Kingdom

91335471R00230